Carnegie Learning®

INTEGRATED MATH I

Assignments and Assessments

FOREWORD TO THE TEACHER

Carnegie Learning's Cognitive Tutor curriculum is dedicated to the teachers, administrators, researchers, friends, family, and students who have made it possible. It continues to be reworked and revised as we strive to offer the best mathematics solutions.

The new textbook format provides a classroom learning environment that parallels the design of the computer portion of the Cognitive Tutor program. This sleek new look provides greater accessibility to mathematics content for both students and teachers through a form that mirrors the Windows® interface used in the software.

Carnegie Learning continues to provide supportive guidance for students in the development of new mathematics concepts, and to provide increased support to teachers for their individual classroom implementations of the Carnegie Learning products.

Students construct mathematical knowledge through the use of the text's new Preview, Input, Process, Analysis and Extension activities in each lesson. As always, the Cognitive Tutor curricula's instructional approach centers on the use of multiple representations, collaborative learning and communication through writing and oral presentations. Our new text design serves to reinforce Carnegie Learning's commitment to meeting educational standards.

Have fun Learning by Doing!

William S. Hadley,
Author, Vice President of Educational Services and Chief Academic Officer
Carnegie Learning, Inc.

P.S. Have questions about the curriculum? Ask me directly by sending an email to billhelp@carnegielearning.com. I will be happy to send you a response within 5 to 7 days. I look forward to hearing from you!

ISBN 1-930804-96-2
Assignments and Assessments

Printed in the United States of America.
1-2003-VH
2-2005-VH

Carnegie Learning, Inc. | Pittsburgh, PA
Phone 888.851.7094 | Fax 412.690.2444
www.carnegielearning.com

Contents

Assignments

Unit 1: Organizing Single Variable Data

Unit 2: Patterns and Linear Functions

Unit 3: Proportional Reasoning

Unit 4: Modeling Situations Using Multiple Representations

Unit 5: Linear Equations: Solving, Graphing, Intercepts, and Slope

Unit 6: Area and Perimeter

Unit 7: The Geometry and Algebra of Right Triangles

Unit 8: Quadratics

Unit 9: Statistical Analysis

Assignment #1 for Bar Graphs, Line Plots and Dot Plots

Name ______________________________

Invent a problem situation that is best displayed using a line graph.

Describe to a reader the type of information you may receive by reading the graph.

Answers will vary but should reflect the following guidelines:

The problem must represent a relationship in which one quantity is dependent upon the other.

The reader should see the relationship between the quantities and be able to describe it, especially if time, dollars or some other numeric measure is involved in the change (whether increasing or decreasing).

Students should choose the appropriate axes to represent the change and make sure that the axes are labeled.

Data sets should be small, perhaps between 10 and 20 data points.

Depending on the relationship, it might be possible to gather historical or current information as well as predict outcomes.

Assignment #2 for Bar Graphs, Line Plots and Dot Plots

Name______________________________

Invent a problem situation that may be described by a dot plot.

Describe to a reader how to interpret the information provided.

Answers will vary but should reflect the following guidelines:

Students will need to generate a situation in which data points have a label attached to each point. Survey-type questions are a likely example.

The information in this case would be the number or percent of respondents.

Interpretation should focus on being able to tell the largest and smallest values and the fact that it is not possible to tell if there are data clusters or gaps.

Assignment #1 for Stem and Leaf Plots

Name ______________________________

You have a theory that foods that are higher in protein are lower in carbohydrates or vice versa. To test out this theory, you will need to gather data by looking at food labels. You will probably want to look at no more than 25 different types of foods in each category.

Using a stem and leaf plot or a split stem and leaf plot (where you organize the data on each side of the stem), compare the distributions and justify your conclusion.

Answers will vary but should reflect the following guidelines:

Students should have examined 20 – 25 products to have a reasonable display in order to make the comparison.

In assessing students' work, see if they find a relationship, direct or indirect, between protein and carbohydrate content. If there is no relationship, see if they indicate why—perhaps there are an insufficient number of data points.

Assignment #2 for Stem and Leaf Plots

Name ______________________________

You have a theory that foods that are higher in fat are lower in sugar or vice versa. To test out this theory, you will need to gather data by looking at food labels. You will probably want to look at no more than 25 different types of foods in each category.

Using a stem and leaf plot or a split stem and leaf plot (where you organize the data on each side of the stem), compare the distributions and justify your conclusion.

Answers will vary but should reflect the following guidelines:

As in Assignment #1, students should examine 20 – 25 products in order to have a reasonable display for the comparison.

In assessing students' work, see if they find a relationship between sugar and fat content. If no relationship is found, see if they indicate why—perhaps there are an insufficient number of data points.

Assignment #1 for Histograms

Name ______________________________

Using the example of the number line plot in your text, describe how you might represent the data using a histogram. Then draw the display.

Which is most effective for determining the distribution of heights in your class?

Sample Response:

We can represent the data from the number line plot as a histogram by placing the heights on the x-axis and the frequency on the y-axis. We can group the heights by inches rather than half inches.

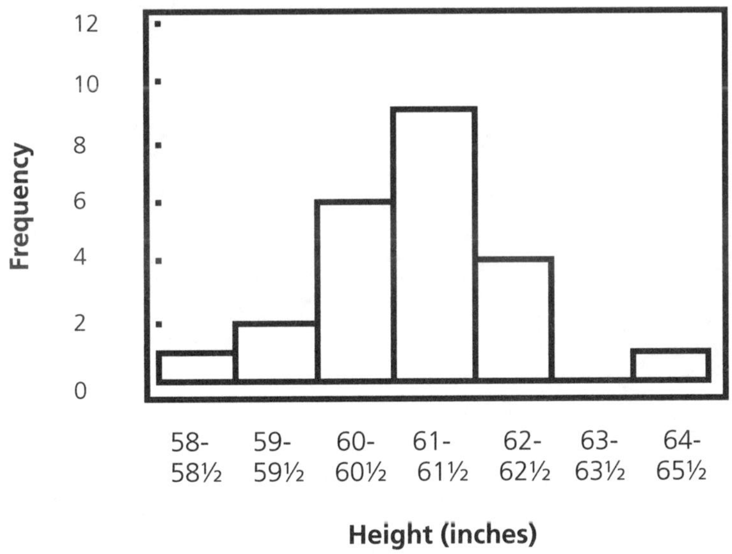

Assignment #2 for Histograms

Name ______________________________

The table below indicates the number of US House Representatives by state.

State	Number of Reps	State	Number of Reps
Alabama	7	Montana	1
Alaska	1	Nebraska	3
Arizona	6	Nevada	2
Arkansas	4	New Hampshire	2
California	52	New Jersey	13
Colorado	6	New Mexico	3
Connecticut	6	New York	31
Delaware	1	North Carolina	12
Florida	23	North Dakota	1
Georgia	11	Ohio	19
Hawaii	2	Oklahoma	6
Idaho	2	Oregon	5
Illinois	10	Pennsylvania	21
Indiana	5	Rhode Island	2
Iowa	5	South Carolina	6
Kansas	4	South Dakota	1
Kentucky	6	Tennessee	9
Louisiana	7	Texas	30
Maine	2	Utah	3
Maryland	8	Vermont	1
Massachusetts	10	Virginia	11
Michigan	16	Washington	9
Minnesota	8	West Virginia	3
Mississippi	5	Wisconsin	9
Missouri	9	Wyoming	1

Source: New York Times Almanac 2001

Represent the data using a histogram.

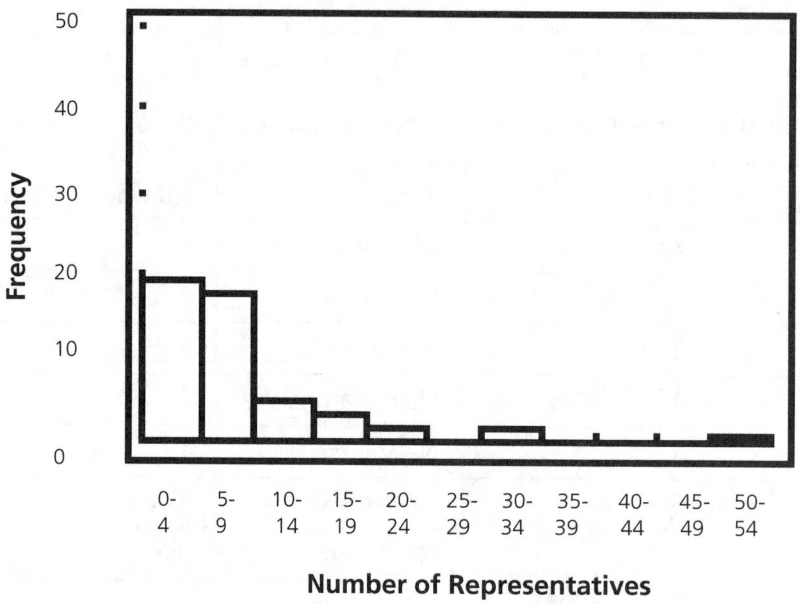

Describe the distribution.

The highest frequency of representatives is between 0 and 14. There is one state with significantly more representatives than the others.

What states have the highest frequency?

California, Florida and New York are the states with the highest frequencies.

Are there outliers?

Yes, there are outliers.

If so, which states?

California is an outlier.

Assignment for Left-Handed Learners

Name: ____________________________

1. Even after performing this experiment in class, you are just not satisfied. That evening, you log onto the internet and find that approximately 12% of the population is left-handed. *Assuming* this is an accurate fact, how many left-handed people should there be in your math class? in your school? your city? state? How many left-handed people live in the United States? If the world population is approximately 6.1 billion, how many people are left-handed?

Number of left-handed people = 12% of (Total Population of class, school, city or state)

Worldwide = 732,000,000 people are left handed

2. Your teacher volunteers that when she was in high school, her class did the same type of survey and found that 17 of the 250 students in the graduating class were left-handed. The rest were right-handed. Using only the data from your teacher's high school class and from your current class, has the number of left-handed students increased? decreased? remained the same? If a change had occurred, do you think this change is "significant"? Explain.

$$17 = x\%(250)$$

$$\frac{17}{250} = x\%$$

$$.068 = x\%$$

$$6.8 = x$$

The percent of left-handed people is almost double today; this statistic fits with the fact that innately left-handed individuals are not being forced to use their right hand. Since the percent is almost double, this seems to be a significant change.

Assignment for $8 an Hour Problem

Name: ______________________________

You are working as a math tutor at a learning center, making $8 an hour. Answer the following questions in complete sentences.

1. If you work 5 hours each day, how much money will you make in:

    a. one week (5 days)?

    The amount earned for five hours per day for five days at $8 per hour is $200.

    b. one month (20 days)?

    The amount earned for five hours per day for twenty days at $8 per hour is $800.

    c. one year (240 days)?

    The amount earned for five hours per day for 240 days at $8 per hour is $9,600.

2. How long will it take you to save enough money to buy:

    a. a computer game machine costing $169?

    It will take about 21.1 hours or 4 to 5 days of work to buy a game machine costing $169.

    b. a computer costing $999?

    It will take about 125 hours or 25 days of work to buy a computer costing $999.

    c. a pair of shoes costing $119?

    It will take about 15 hours or 3 days of work to buy a pair of shoes costing $119.

    d. a television costing $249?

    It will take about 31.1 hours or 6 to 7 days of work to buy a television costing $249.

3. If you got a raise of $0.20 per hour, how much *more* money would you make:

   a. each week?

   Earnings would be $5 more per week.

   b. each month?

   Earnings would be $20 more per month.

   c. each year?

   Earnings would be $240 more per year.

Assignment for Number Patterns

Name: ______________________________

For each problem, use complete sentences to answer three questions:

- What do you know about the answer to the problem before you begin?
- What is the solution to the problem? Show your work.
- Does your solution make sense? Why or why not?

1. You have 45 pounds of dog food, and you must divide it among 60 dogs. How many pounds will each dog receive?

The given information is the total number of dogs and the total number of pounds of food. There are more dogs than pounds of food, so the amount will be less than a pound.

Each dog will receive ¾ of a pound.

$$\frac{45}{60} = .75 = \frac{3}{4}$$

The answer makes sense but really depends on the size of the dogs.

2. Two out of every nine teachers are younger than 30. If there are 144 teachers in this building, how many are younger than 30?

Two out of every nine teachers are younger than 30 and there are 144 teachers, so about 1/3 of teachers should be younger than 30.

Thirty-two teachers of the 144 in the building are younger than 30.

$$\frac{2}{9} = \frac{x}{4}$$

$$x = 32$$

The answer makes sense.

3. Only four out of every five baseball players have batting averages *over* 0.200. If there are 750 baseball players, how many have batting averages *less* than 0.200?

   Four out of every five players have batting averages over 0.200 and there are 750 players, so about 80 to 90% of the players will have a batting average of over 200.

   Since 600 players have batting averages over 0.200, 150 players have averages less than 0.200.

$$\frac{4}{5} = \frac{x}{750}$$

$$x = 600$$

$$750 - 600 = 150$$

   The answer makes sense but may not be accurate since it does not account for any players who have a batting average of exactly 0.200.

Assignment for Finding the 10th Term

Name: ______________________________

For each problem, use complete sentences to answer three questions:

- What do you know about the answer to the problem before you begin?
- What is the solution to the problem? Show your work.
- Does your solution make sense? Why or why not?

1. Six partners lost $540 in Las Vegas. If they divide their losses evenly, how much will it cost each partner?

   Since 6 people lost money totaling $540, the amount of loss per person will be less than $100.

   Each partner will lose $90.

$$\frac{540}{6} = 90$$

   The answer makes sense.

2. There are 2,500,000,000 people in the world. If two-thirds of the people in the world live in poverty, how many people live in poverty?

   If there are 2.5 billion people in the world and 2/3 live in poverty, the result should be about 2/3 of 2.5 billion or 16.7 billion people.

   Approximately 1,670,000,000 people live in poverty.

$$\left(\frac{2}{3}\right)(2{,}500{,}000{,}000) \approx 1{,}670{,}000{,}000$$

   The answer makes sense. (Students may make comments on the magnitude.)

3. Half of all the employees in a company have profit sharing. The total profits for the company are \$250,000. If each employee with profit sharing receives an equal share of the profit and there are 1,250 employees in the company, how much will each receive?

   There are 1,250 employees in the company and each shares \$250,000 equally, so 625 people divide up the profits; therefore, the total should be about \$400. Alternatively, if 1,250 employees have profit sharing then each receives \$200.

   Each employee's share of the profit is \$400.

$$\frac{250{,}000}{625} = 400$$

   The answer makes sense.

Assignment for Soaps and Bath Beads

Name: ______________________

1. On one particular Appletree & Elizabeth production line, the ratio of large sized boxes medium sized boxes is 7 to 5. There are exactly 342 bath beads that must be placed i boxes. How many of each type of gift box can be created? Show your work and expl answer.

   There are 12 bath beads in each large box and there are 6 bath beads in medium sized box.

   For every 7 large boxes there are 5 medium boxes, which means that for 84 (12*7) beads in the large boxes, there are 30 (6*5) beads in the med boxes.

   Thus, there are 114 (84+30) beads for every set of 7 large and 5 mediun boxes.

   If the total number of available beads is 342, then there must be 3 sets c large and 5 medium boxes. (342/114 = 3)

   This means there are 21 (7*3) large boxes, and 15 (5*3) medium boxes.

2. How many scented soaps are needed on the production line to fill the boxes? Make s show your work and explain your answer.

   There are 6 soaps in every large box. If there are 21 large boxes, there a (21*6) soaps in the large boxes.

   There are 2 soaps in every medium box. If there are 15 medium boxes, t 30 (15*2) soaps in the medium boxes.

   This makes a total of 156 (126 + 30) soaps altogether in the 21 large bo 15 medium boxes.

Assignment for The Consultant Problem

Name: ____________________________

1. A friend of your sister consults as a landscape designer and charges $67.50 per hour.

   Complete this table.

	Hours Worked	Earnings
Labels	Hours Worked	Earnings
Units	hours	dollars
	1	***67.50***
	2	***135.00***
	5	***337.50***
	8	***540.00***
	10	***675.00***
	11	***742.50***
	15	***1,012.50***
	18	***1,215.00***
	20	***1,350.00***

Would a line graph be useful to illustrate the information in this table? Why or why not?

Yes because it would provide a sense of the rate at which the amount earned increases with the number of hours worked. It would also indicate the amount earned for any number of hours based on the upper and lower bounds.

2. Construct a line graph on this grid. Include all the information from the table. Label the graph clearly. Your teacher will give your group an overlay, so that the whole class can see your graph.

Use your graph to answer the following questions. Write your answers in complete sentences.

3. How much will your sister's friend make if he works:

    a. 10 hours? ***He will earn $675.00 in 10 hours.***

    b. 22 hours? ***He will earn $1485.00 in 22 hours.***

Write a sentence describing how you found these answers.

Sample: I found the number of hours worked on the horizontal axis. I traced this coordinate up to find the corresponding y-coordinate on the line. The y-coordinate tells the amount of money earned.

4. How much will your sister's friend make if he works:

   a. 3 hours 30 minutes? ***Your sister's friend will make $ 236.25. (3.5*67.5)***

   b. 5 hours 12 minutes? ***Your sister's friend will make $351.00. (5.2*67.5)***

   c. 7 hours 45 minutes? ***Your sister's friend will earn $523.13.***

   d. 8 hours 40 minutes? ***Your sister's friend will earn $585.23.***

   Write a sentence describing how you found these answers.

   First, the minutes must be converted to a part of an hour. Then, multiply the number of hours worked by the hourly rate, which is $67.50.

5. How many hours will he have to work to make:

   a. $1,000? ***Your sister's friend will have to work 14.81 hours to make $1000.***

   b. $500? ***He will have to work 7.41 hours to make $500.***

   c. $33,000? ***He will have to work 488.89 hours to earn $33,000.***

   Write a sentence describing how you found these answers.

   Divide the amount of money your sister's friend will earn by the hourly wage of $67.50 to calculate the number of hours he will need to work.

6. Remember that mathematicians spend a great deal of time looking for number patterns. Use complete sentences to describe the number pattern in your table.

   The pattern shows that for each hour your sister's friend works, he earns $67.50 more. When the numbers increase by 1 in the hours column they will increase by 67.5 in the earnings column.

7. If your sister's friend worked *H* hours, what is the expression for finding his total earnings?

   $67.50 * H

Assignment for U.S. Shirts

Name: ______________________________

You are going on vacation by car. You will be traveling at an average speed of 50 miles per hour.

Make a table showing how far you will travel in various time periods from 1 to 20 hours. Use this information to construct a line graph and find the algebraic formula.

Labels	Hours Driven	Total Miles
Units	hours	miles
	1	***50***
	2	***100***
	5	***250***
	11	***550***
	15	***750***
	18	***900***
	20	***1000***

Algebraic Expression: ***50T***

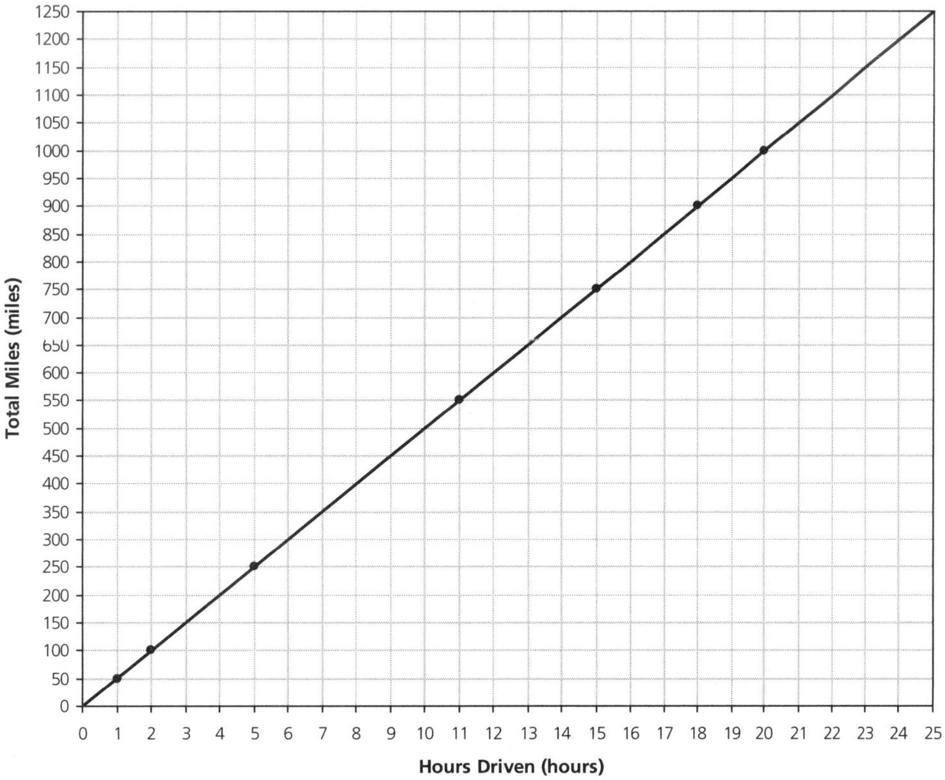

Assignment for Hot Shirts

Name: ______________________________

You are going on vacation by airplane. You will be traveling at an average speed of 325 miles per hour.

Make a table showing how far you will travel in various time periods from 1 to 20 hours. Use this information to construct a line graph and find the algebraic formula.

Labels	Time in the Air	Distance Traveled
Units	***hours***	***miles***
Formula	***t***	***325t***
	1	***325***
	2	***650***
	5	***1625***
	11	***3575***
	15	***4875***
	18	***5850***
	20	***6500***

Algebraic Expression: ***325T***

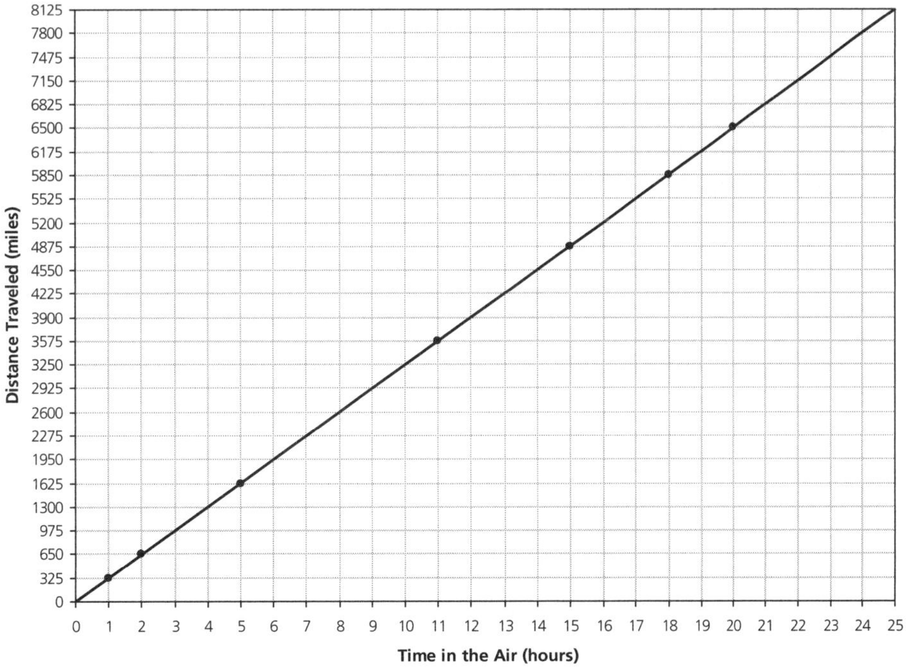

Assignment for Comparing U.S. and Hot Shirts

Name: ______________________________

A friend is employed at a company where she makes $9 an hour.

Complete this table, including the algebraic formula. Draw a line graph displaying her total earnings and clearly label the axes.

Labels	***Number of Hours***	***Total Earnings***
Units	***Hours***	***Dollars***
Formula	***H***	***9H***
	1	***9***
	2	***18***
	5	***45***
	11	***99***
	15	***135***
	18	***162***
	20	***180***

Algebraic Expression: ***9H***

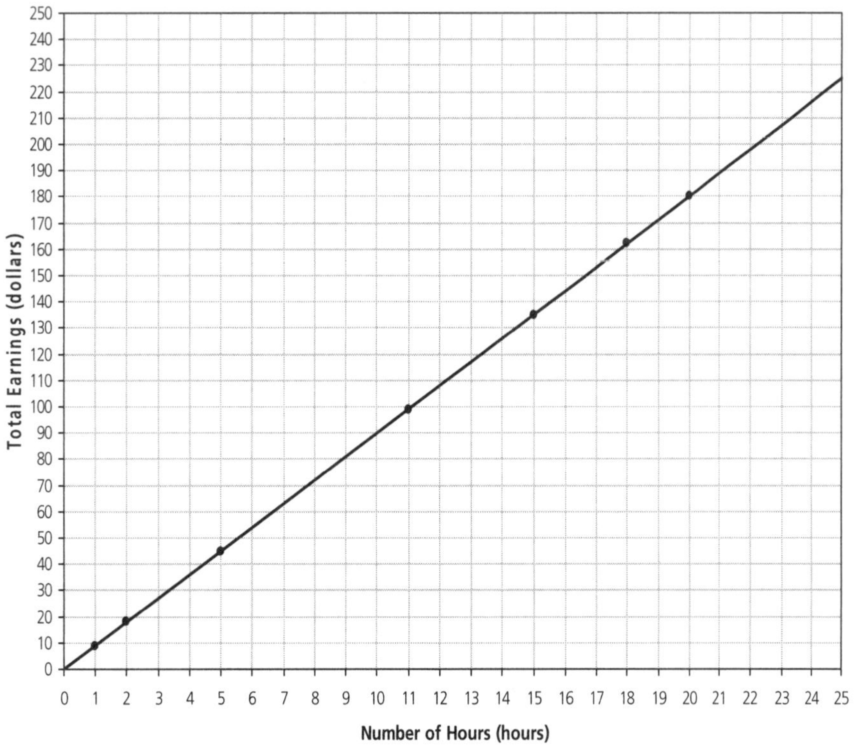

Assignment for Finding the Nth Term

Name: ____________________________

For each problem, use complete sentences to answer three questions:

- What do you know about the answer to the problem before you begin?
- What is the solution to the problem? Show your work.
- Does your solution make sense? Why or why not?

1. Five partners made a profit of $2,540 on their enterprise. If the profits are to be divided evenly, how much will each partner receive?

   Before beginning the problem, we know there are five people dividing a total of $2540, so that Jack will get about $500.

   Each partner receives a total of $508.   2540/5 = $508.

   The solutions do make sense because it is very close to the estimated answer.

2. There are 250 million people in the United States. If one out of every ten people has an annual income *over* $60,000, how many people make *less* than $60,000 per year?

   If there are 250 million people and 1 out of every 10 has an income over $60,000, then 9 out of 10 have incomes under $60,000.

   $$\frac{9}{10} = \frac{x}{250000000}$$

   x = 225,000,000 people will incomes under $60,000.

   The answer does make sense, but it includes all of the people who are too young to work, so it is inflated somewhat.

3. Three out of every five students are in a certain club, and each club member is to receive an equal share of the profits. If there are 1,000 students and the profits total $3,600, how much will each club member receive?

   If three of every five are in a club and there are 1000 students total, then about 600 are in the club. If $3600 is divided among 600 people, each person will get a small amount.

   $3600/600 = $6

   The answer makes sense for the many people who are in the club.

Assignment for Handshake Problem

Name: ______________________________

Teams in a League Problem

You have been asked to help schedule the games in your younger sister's soccer league. There are 8 teams in the league. You have been asked to determine the total number of games that must be played so that *every team will play every other team exactly once*.

Someone suggests that to find the answer, you first try to figure out the total number of games for a league that has three teams, then four teams, and so on.

You must make a written presentation to the league officers. The report should include a description of how you arrived at your answer. Write the report below.

Students should use the strategy they used in the Handshake Problem. They may also use simpler problems as suggested in the directions, using 3 teams in the league, then 4 teams and so on up to 8 teams.

If n=8, then the number of games will be 7+6+5+4+3+2+1+0 = 28.

Another strategy students might use is to list the games. For example:

team 1 plays team 2

team 1 plays 3

1 plays 4

etc...

Student might also create a table, grid or picture to connect the teams as they play each other.

No matter the strategy used, the final written report should include the total number of games as 28. The final report should also include the description of whatever strategy the student used to deduce the answer of 28 games.

Assignment for Gauss' Solution

Name: ______________________________

For each problem, perform the following three tasks:

- Write something you know about the answer before you solve the problem. Highlight or underline the important information.
- Solve the problem. Show all of your work.
- Write your answer in a complete sentence.

1. **One out of every twenty students** is on the basketball team. If there are **1,000 students** in the school, how many are on the basketball team?

   The important information is in bold in the teacher's edition only

   Before solving the problem, it is known that the number of students on the team will be less than 100 because only 1 out of 20 is on the team.

   1000/20 = 50

   There will be 50 people on the basketball team.

2. A recent advertisement offered a **25% discount** on athletic shoes. If the **regular price is $60**, what is the sale price?

   Before solving the problem, it is known that the price of the shoes is less than $60.

   0.25 * 60 = 15

   60 – 15 = 45

   The shoes will cost $45 on sale.

3. A recent census reported that **10%** of the people in the United States have **incomes over $70,000** and **30%** of the people have **incomes less than $10,000.** If there are **250 million people** in the United States, how many are included in each of these two groups?

   Before solving the problem, it is known that the number of people with incomes about $70,000 is less than the number of people with incomes under $10,000.

   10% of 250 million = 25 million

   30% of 250 million = 75 million

   There are 25 million people with incomes above $70,000,and 75 million with incomes under $10,000.

Assignment for Shadows and Proportions

Name: ______________________________

In the movie, *Honey, I Shrunk the Kids,* Rick Moranis is a hair-brained scientist who has created a shrinking machine. When the machine accidentally gets activated, his kids get shrunk.

In the movie, there are a number of shots of the kids in a variety of predicaments.

In one shot, a child appears to be $\frac{1}{2}$ the length of the father's nose, while the child climbs onto the father's head.

In another shot, the same child's arm appears to be about $\frac{1}{6}$ the diameter of a Cheerio while the child is floating in a bowl with cereal and milk.

Are these two pictures consistent if the original height of the child is approximately 4 feet and the child's arm is about 2 feet long?

Fully analyze this situation and justify any conclusions you make about the shrunken child.

An adult nose is about 2 inches, which means the child is about 1 inch tall.

The diameter of a Cheerio is about ½ inch, so the child's arm is about $\frac{1}{12}$ ***inch.***

These two pictures are not consistent. If these pictures were consistent, the child's arm would be about ½ inch, not $\frac{1}{12}$ ***inch.***

In the movie promotion, the kids were supposedly shrunken to only $\frac{1}{4}$ inch tall. Does that agree with your calculations above?

This does not agree with either of the two calculations in the above problem.

Problem adapted from *Measurement in the Middle Grades*, NCTM, 1994

Assignment for Making Punch

Name: ____________________________

1. A typical male lion is about 2.1 meters long and stands about 90 cm tall at the shoulder. He also weighs about 240 kg. The Sphinx, located near Cairo, Egypt, is a depiction of a man – lion combination. The Sphinx is about 73.2 meters long. This includes the paws, which are 15.3 meters long. If there existed a real lion that were the size of the Sphinx, how tall would you expect it to be? Show your work and explain any assumptions you made.

   Assuming the 2.1 meters as the length of a lion does not include the paws, then the Sphinx is 57.9 meters.

   $$\frac{2.1 \text{ meters}}{0.9 \text{ meters}} = \frac{57.9 \text{ meters}}{x \text{ meters}}$$

   x = 24.8 meters

   If the lion were the size of the Sphinx, it would be 24.8 meters tall

2. If you make the assumption that the Sphinx's paws are in proportion to the rest of its body, how long is the paw on an actual male lion?

   15.3 meters is about $\frac{1}{5}$ ***of 73.2 meters, so the lion's paw should be about*** $\frac{1}{5}$ ***of the size of its body.***

   $\frac{1}{5}$ ***of 2.1 meters = 0.42 meters or 42 centimeters***

3. **Challenge:** If the Sphinx were a real lion, how much would you expect it to weigh? (Hint: Is weight dependent only upon height?)

Weight is dependent upon all three dimensions of the lion, so the ratio of weights between the Sphinx and the lion is the cube of the ratio of lengths

$$\frac{57.9}{2.1} = 27.57$$

27.57^3 = 20,956.09, so the weight of the lion must be multiplied by 20,956 to calculate the weight of the Sphinx.

240*20,956 is about 5 million pounds.

Adapted from *World's Largest Math Event 4*

Assignment for TV News Ratings

Name: ____________________________

Two out of every three four-year-olds like crackers.

Complete this table, including the algebraic formula. Draw a line graph illustrating this situation and clearly label the axes.

Labels	Total 4-year-olds	4-year-olds Who Like Crackers
Units	***people***	***people***
Formula	***T***	***(2/3)T***
	3	***2***
	30	***20***
	300	***200***
	3000	***2000***
	30,000	***20,000***
	300,000	***200,000***

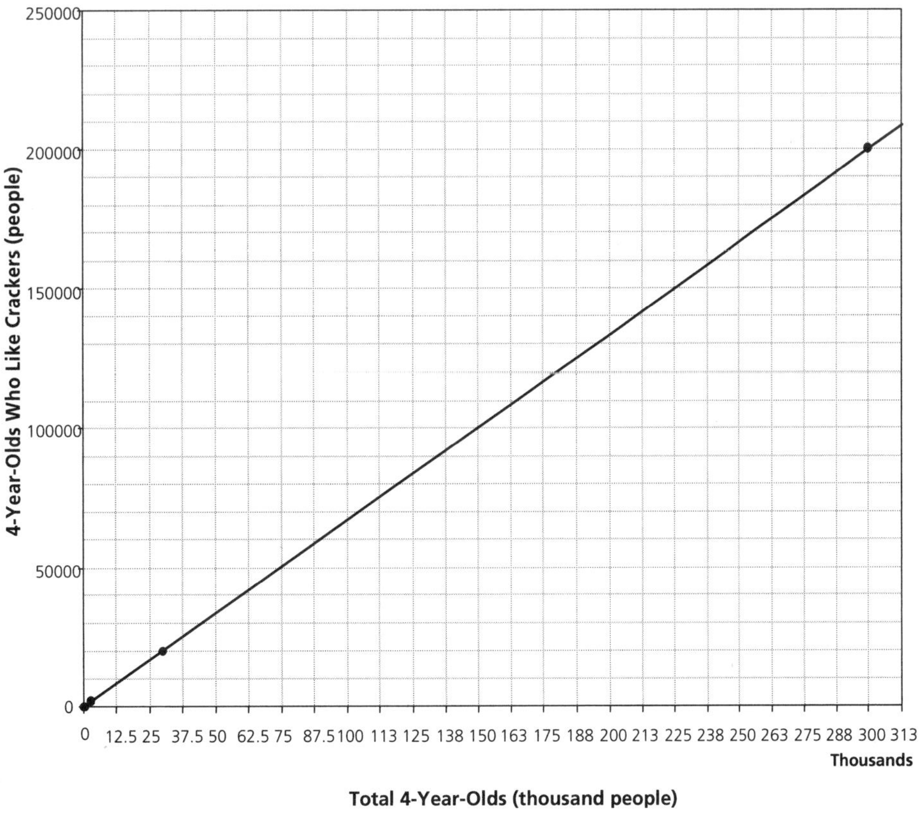

Assignment for Truck Life

Name: ______________________________

Three out of four algebra students will receive an A, B, or C.

Complete this table, including the algebraic formula. Draw a line graph illustrating this situation and clearly label the axes.

Labels	Total Algebra Students	Those Who Receive an A, B, or C
Units	***People***	***People***
Formula	***P***	***(3/4)P***
	100	***75***
	200	***150***
	300	***225***
	400	***300***
	500	***375***
	600	***450***

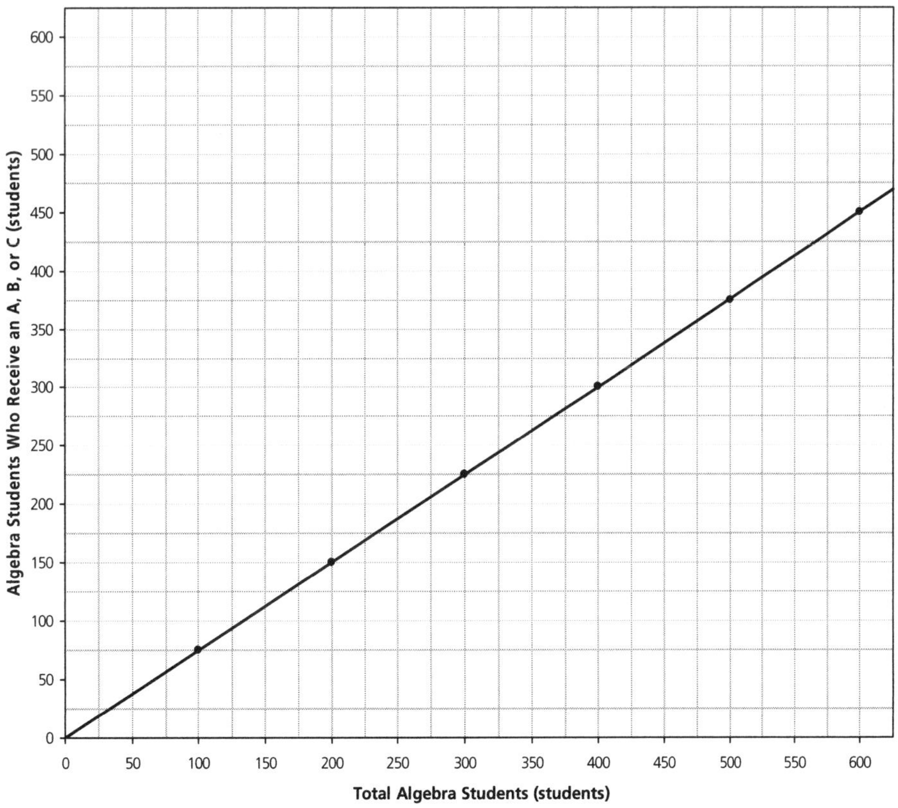

Assignment for Women at a University

Name: ______________________________

One out of every fifty drivers will have an automobile accident this year.

Complete this table, including the algebraic formula. Draw a line graph illustrating this situation and clearly label the axes.

Labels	Total Drivers	***Number of Drivers Who Have Had an Accident***
Units	***People***	***People***
Formula	***P***	***(1/50)P***
	1,000,000	***20,000***
	5,000,000	***100,000***
	10,000,000	***200,000***
	75,000,000	***1,500,000***
	100,000,000	***2,000,000***
	150,000,000	***3,000,000***

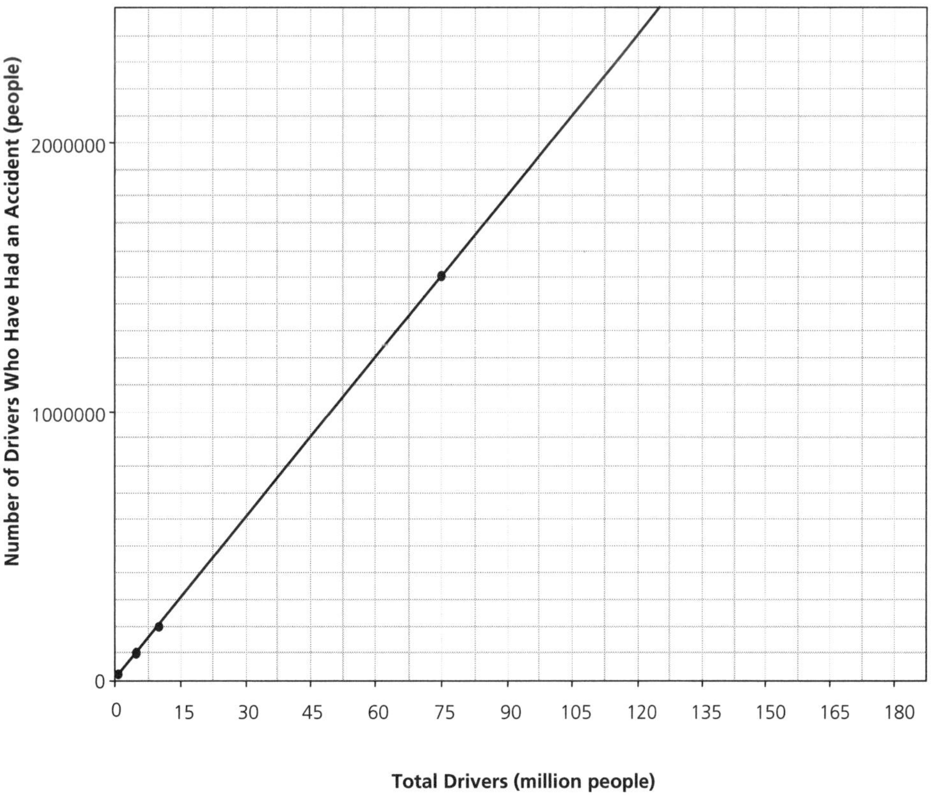

Assignment for Plastic Containers

Name______________________________

The force of gravity on the surface of the Moon is about $\frac{1}{6}$ of the force of gravity on the Earth's surface.

1. How high can you jump vertically? (If you do not know, estimate by standing next to a wall and jumping straight up and touching the wall at the top of your jump.)

   Answers will vary.

2. If you were on the surface of the Moon, how high could you jump? Show your work and explain what you did to solve.

   Answers will vary, but should be 6 times the answers from question number 1.

3. How far can you long jump (with a running start and without a running start)? (If you don't know, mark a starting point on the floor and jump as far as you can.)

   Answers will vary from about 4 feet for a standing long jump to much farther for one with a running start.

4. How far could you long jump on the surface of the moon – with and without a running start?

   Answers will vary and should be 6 times as far as the previous two answers given.

5. In track and field, hurdles are about 0.91 meters high for women and 1.07 meters. In order to make the event fair, how high would the hurdles have to be on the Moon?

   The hurdles wold have to be 6 times as high.

   (0.91)*(6) = 5.46 meters for women and

   (1.07)*(6) = 6.42 meters high for men.

Adapted from *World's Largest Math Event 3*

Assignment for Mia is Growing Like a Weed

Name ______________________________

The typical gestation period (time from conception to birth) for a human baby is about 40 weeks (a bit longer than 9 months). Recently, the advanced development of ultrasound scanning of fetuses has begun to include the ability to make measurements of various parts of the baby's body. By evaluating large groups of fetuses that later had normal birth weights, it has been possible to create standard tables of fetal growth. The table below contains data about the length of the femur of a fetus at different times throughout the gestation period.

Quantity Name	**Gestation Time**	**Femur Length**
Units	weeks	mm
	12	10
	12.5	12
	13	13
	15	14
	20	15.5
	25	16.9
	30	18.8
	40	21

Create a scatter plot of this data, making sure to label the axes appropriately and making sure to use an appropriate scale.

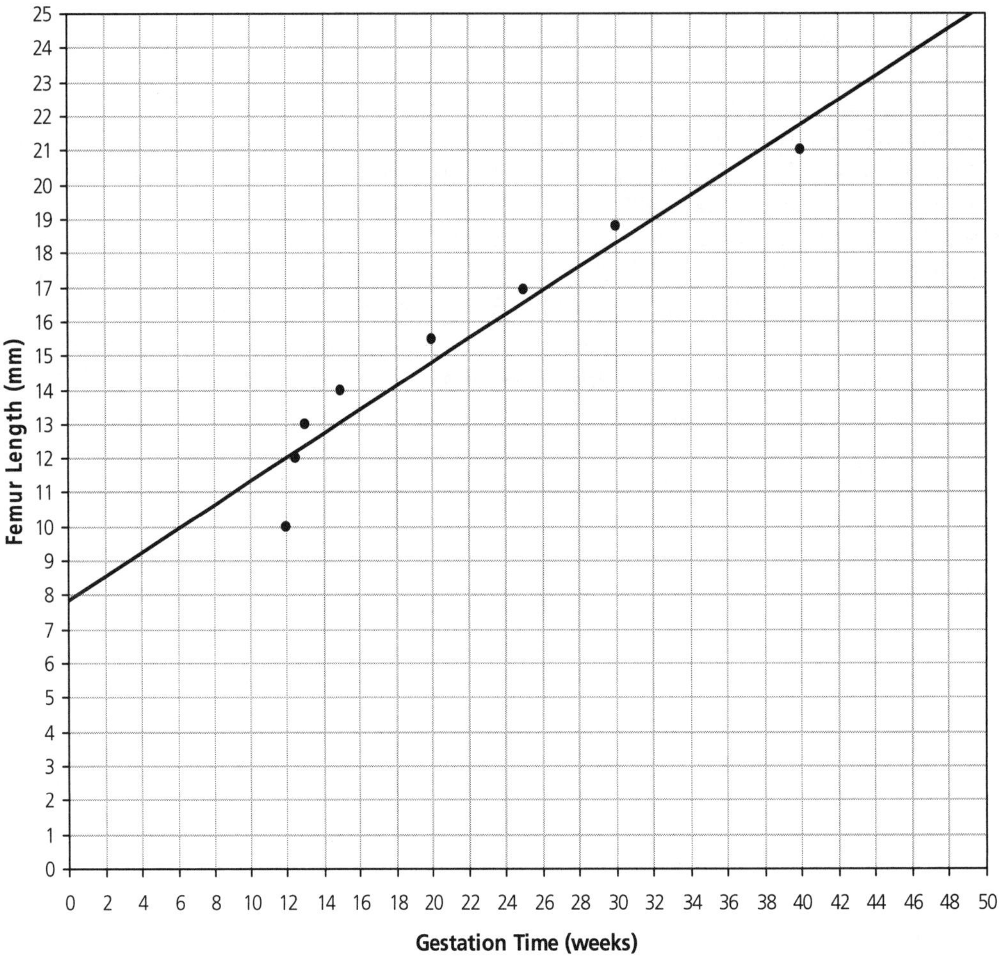

1. Create a line that best approximates the data (your line of best fit).
2. Approximately how many millimeters is the femur growing every week? Explain how you know.

   Sample: The femur is growing about ½ of a millimeter each week. In 20 weeks the femur grows to 21 mm, so that is close to 0.5 mm each week. Also, the line of best fit comes close to (40, 20).

3. If the femur of the fetus continues to grow at this same rate, how long will it be when the baby is 3 months old?

   Assuming 3 months is 12 week, the additional growth will be 6 mm. If the baby was born with a 21 mm femur, it will be 27 mm long at 3 months of age.

4. Is it reasonable that the femur of the baby will continue to grow at the same rate as before the baby was born? Why or why not?

   It is not reasonable that this rate of growth would continue over a number of years. The growth rate would have to slow after a period of time.

Assignment for Taxes Deducted from Your Paycheck

Name:____________________________

In Pennsylvania there is a 6% sales tax on the total price of items sold.

Complete this table, including the algebraic formula. Draw a line graph illustrating this situation and clearly label the axes.

Sample Answer:

Labels	Total Price	Sales Tax
Units	***Dollars***	***Dollars***
Formula	***C***	***.06(C)***
	10	***.60***
	15	***.90***
	25	***1.50***
	37	***2.22***
	42	***2.52***
	86	***5.16***
	120	***7.20***

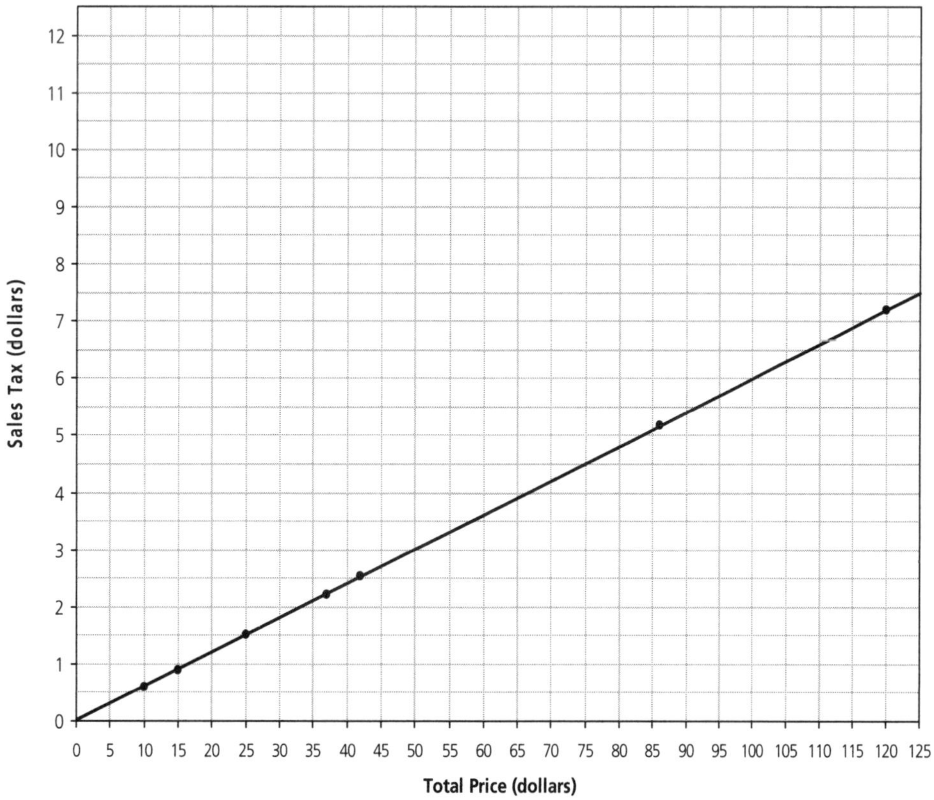

Assignment for Tipping in a Restaurant

Name: ______________________________

Approximately 76% of ninth-graders in the Pittsburgh Public Schools will graduate from high school in four years.

Complete this table, including the algebraic formula. Draw a line graph illustrating this situation and clearly label the axes.

Sample Answer:

Labels	Ninth-Graders	Those Who Graduate
Units	***People***	***People***
Formula	***G***	***.76G***
	350	***266***
	600	***456***
	900	***684***
	1500	***1140***
	2500	***1900***
	3000	***2280***

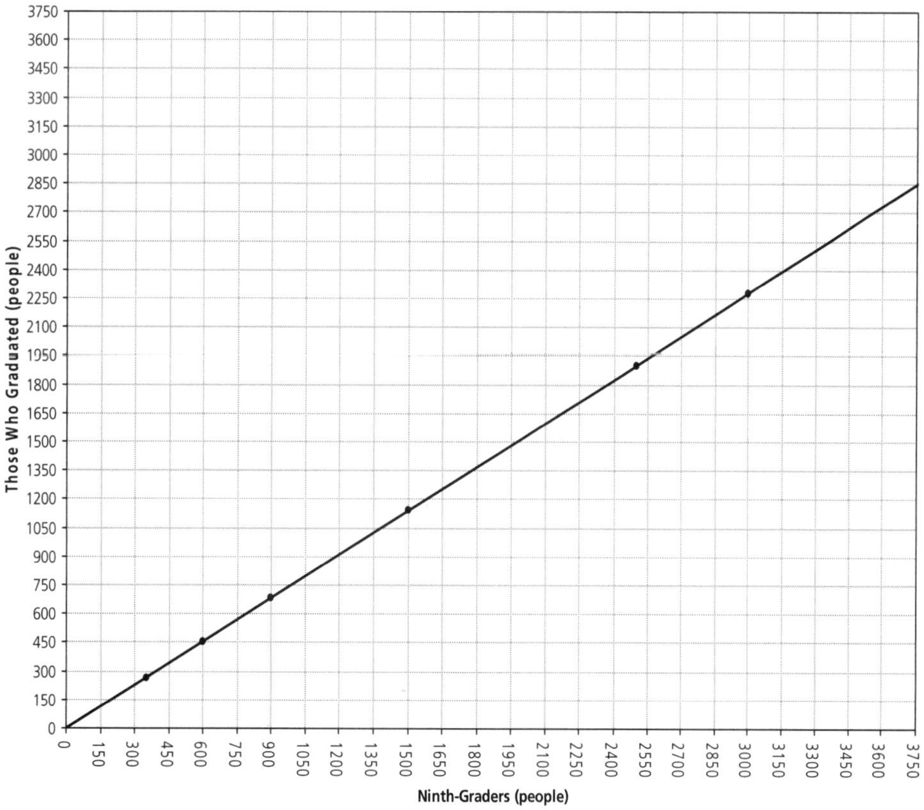

Assignment #1 for Earning Sales Commission

Name: ______________________________

A television salesperson is paid $150 per week, plus a commission of 5% on total sales.

Complete this table, including the algebraic formula. Draw a line graph illustrating this situation and clearly label the axes.

Labels	***Total Sales***	***Total Earnings***
Units	***Dollars***	***Dollars***
Formula	***S***	***150 + .05S***
	500	***175***
	800	***190***
	1000	***200***
	3000	***300***
	7500	***525***
	10000	***650***

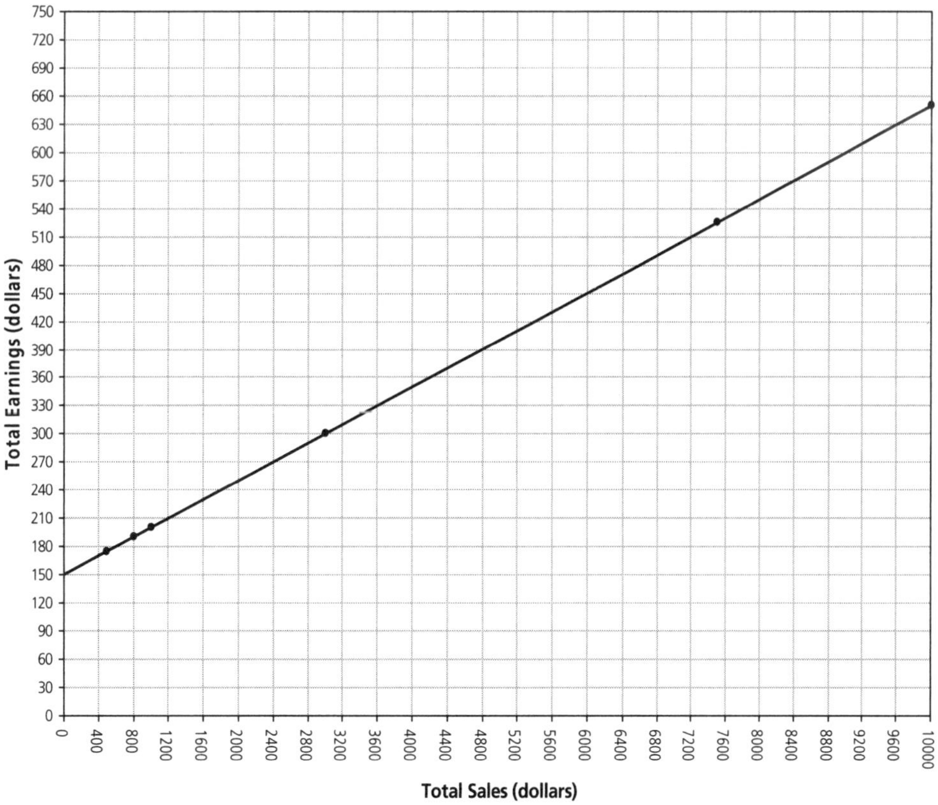

Assignment #2 for Earning Sales Commission

Name: ______________________________

A television salesperson is paid a commission of 5% on total weekly sales plus a $500 per week base salary.

Complete this table, including the algebraic formula. Draw a line graph illustrating this situation and clearly label the axes.

Labels	***Total Weekly Sales***	***Total Earnings***
Units	***Dollars***	***Dollars***
Formula	***S***	***500 + 0.05S***
	620	***531***
	780	***539***
	1000	***550***
	1240	***562***
	1460	***573***
	2000	***600***

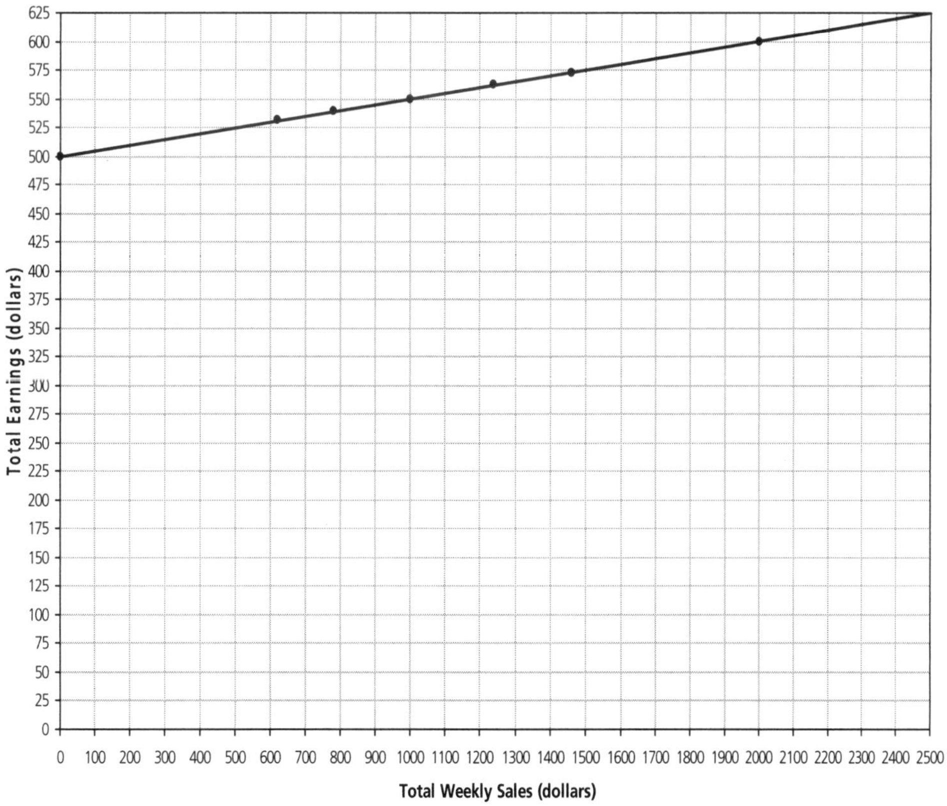

Assignment for Rent a Car from Go-Go Car Rentals

Name: ____________________________

A computer rents for $2.50 per day, with a $25 fee for insurance.

1. Find the total cost of renting the computer for:

   a. 20 days

   The total cost of renting the computer for 20 days is $75.00.

   b. 33 days

   The total cost of renting the computer for 33 days is $107.50.

   c. Write a complete sentence describing how you found these answers.

   Multiply the number of days by the cost per day and then add the insurance cost.

2. Write an algebraic equation for the total cost of renting a computer.

   Total rent = ***2.5x + 25***

3. For how many days can you rent the computer for $67.50?

   a. Write an equation that states that the rental cost is $67.50.

   2.50 x + 25 = 67.50

   b. Solve this equation, showing your work.

   2.50 x = 42.5

   x = 17

   c. Write your answer to question 3 in a complete sentence.

   You can rent the computer for 17 days for a cost of $67.50.

   d. Write a complete sentence describing how you solved the equation.

   Subtract the insurance cost from the rental cost and then divide by the cost per day.

4. For how many days can you rent the computer for \$202.50?

   a. Write an equation that states that the rental cost is \$202.50.

   2.50 x + 25 = 202.50

   b. Solve this equation, showing your work.

   2.5 x = 177.5

   x = 71

   c. Write your answer to question 4 in a complete sentence.

   You can rent the computer for 71 days for a cost of \$202.50.

   d. Write a sentence describing how you solved the equation.

   Subtract the insurance cost from the total rental cost. Then divide by the cost per day.

5. For how many days can you rent the computer for \$92.50?

   a. Write an equation that states that the rental cost is \$92.50.

   2.50 x + 25 = 92.50

   b. Solve this equation, showing your work.

   2.5 x = 67.5

   x = 27

   c. Write your answer to question 5 in a complete sentence.

   You can rent the computer for 27 days for a cost of \$92.50.

   d. Write a sentence describing how you solved the equation.

   Subtract the insurance cost from the rental cost. Divide the result by the cost per day.

6. Use the information from questions 1–5 to complete this table.

Labels	Time	Total Rental Fee
Units	***days***	***dollars***
Formula	***x***	***2.5 x + 25***
	20	***75.00***
	33	***107.50***
	17	67.50
	71	202.50
	27	92.50

7. Graph the equation for this problem situation.

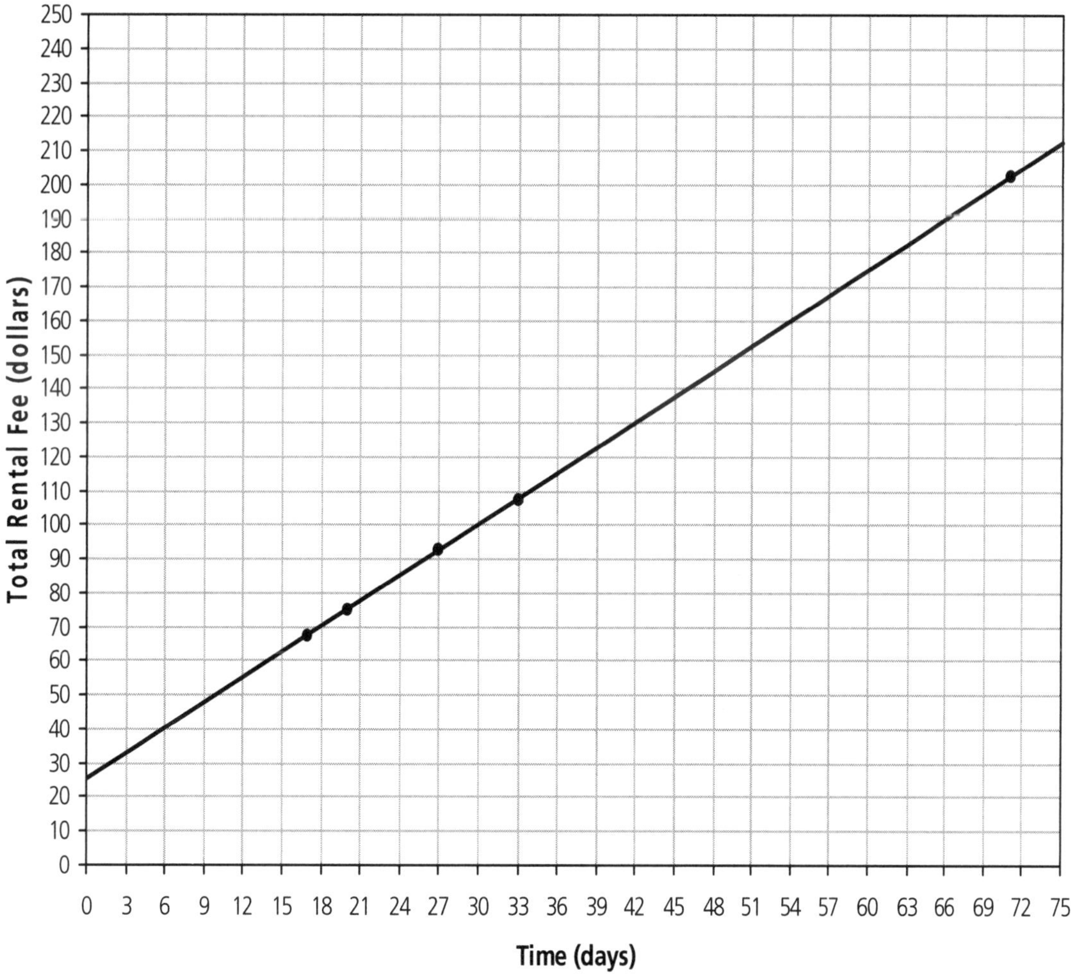

Assignment for Rent a Car from Wreckem Car Rentals

Name: ______________________________

A car rents for $23 per day, plus an insurance fee of $32.

1. What does it cost to rent this car for:

   a. 10 days?

   It costs $262 to rent the car for 10 days.

   b. 37 days?

   The cost to rent the care for 37 days will be $883.

   c. Write a sentence describing how you found these answers.

   Multiply the number of days by the cost per day. Then, add the insurance cost to get the total.

2. Write the algebraic rule for this problem situation.

   Car rental = ***23 x + 32***

3. Write an equation you can use to find out the number of days for which you can rent this car for $331. Solve this equation, showing your work, and write your answer in a complete sentence. Write a complete sentence describing how you solved the equation.

   23 x + 32 = 331

   23 x = 299

   x = 13

   For a cost of $331, you can rent a car for 13 days.

4. Write an equation you can use to find out the number of days for which you can rent this car for \$149. Solve this equation, showing your work, and write your answer in a complete sentence. Write a complete sentence describing how you solved the equation.

   23 x = 32 = 149

   23 x = 117

   x = 5.09

   You can rent the car for 5 days for a little under \$149.

5. Write an equation you can use to find out the number of days for which you can rent this car for \$279. Solve this equation, showing your work, and write your answer in a complete sentence. Write a complete sentence describing how you solved the equation.

   23 x = 32 = 279

   23 x = 247

   x = 10.74

   You can rent a car for 10 days for under \$279.

6. Use the information from questions 1–5 to complete this table.

Labels	***Time***	***Cost***
Units	***days***	***dollars***
Formula	***x***	***23 x + 32***
	10	***262***
	37	***883***
	13	***331***
	5.09	***149.07***
	10.74	***279.02***

7. Graph the equation for this problem situation.

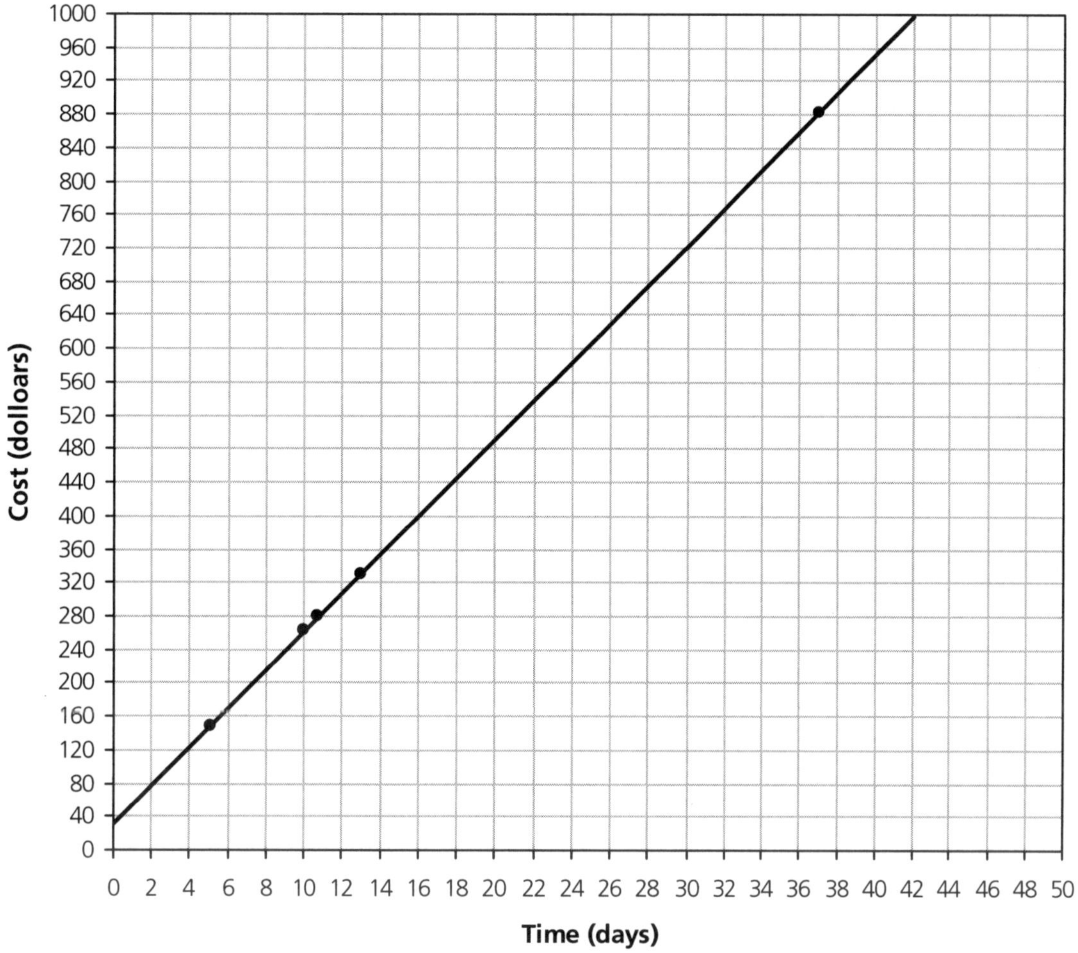

Assignment for Rent a Car from Good Rents Car Rentals

Name: ______________________________

A caterer charges $12.50 per person, plus a $150 fee for the hall rental.

1. Write the algebraic rule for this problem situation.

   Cost of the caterer = ***y = 12.5 x + 150***

2. What would it cost to cater a party for:

   a. 35 people?

   12.5(35) + 150 = 587.5   It will cost $587.50 to cater a party for 35 people.

   b. 97 people?

   12.5(97) + 150 = 1362.5   It will cost $1362.50 to cater a party for 97 people.

3. Write an equation you can use to find out how many people you can invite to a party that costs $687.50. Solve this equation, showing your work, and write your answer in a complete sentence.

   12.5 x = 150 = 687.50

   12.5 x = 537.5

   x = 43

   For a cost of $687.50, you can invite 43 people.

4. Write an equation you can use to find out how many people you can invite to a party that costs $1287.50. Solve this equation, showing your work, and write your answer in a complete sentence.

   12.5 x = 150 = 1286.5

   12.5 x = 1137.5

   x = 91

   For a cost of $1287.50, you can cater a party for 91 people.

5. Write an equation you can use to find out how many people you can invite to a party that costs $2000. Solve this equation, showing your work, and write your answer in a complete sentence.

   12.5 x + 150 = 2000

   12.5 x = 1850

   x = 148

   You can invite 148 people for a cost of $2000.

6. Use the information from questions 1–5 to complete this table.

Labels	***Number of people***	***Cost***
Units	***people***	***$***
Formula	***x***	***12.5x + 150***
	35	***587.50***
	97	***1362.50***
	43	***687.50***
	91	***1287.50***
	148	***2000.00***

7. Graph the equation for this problem situation.

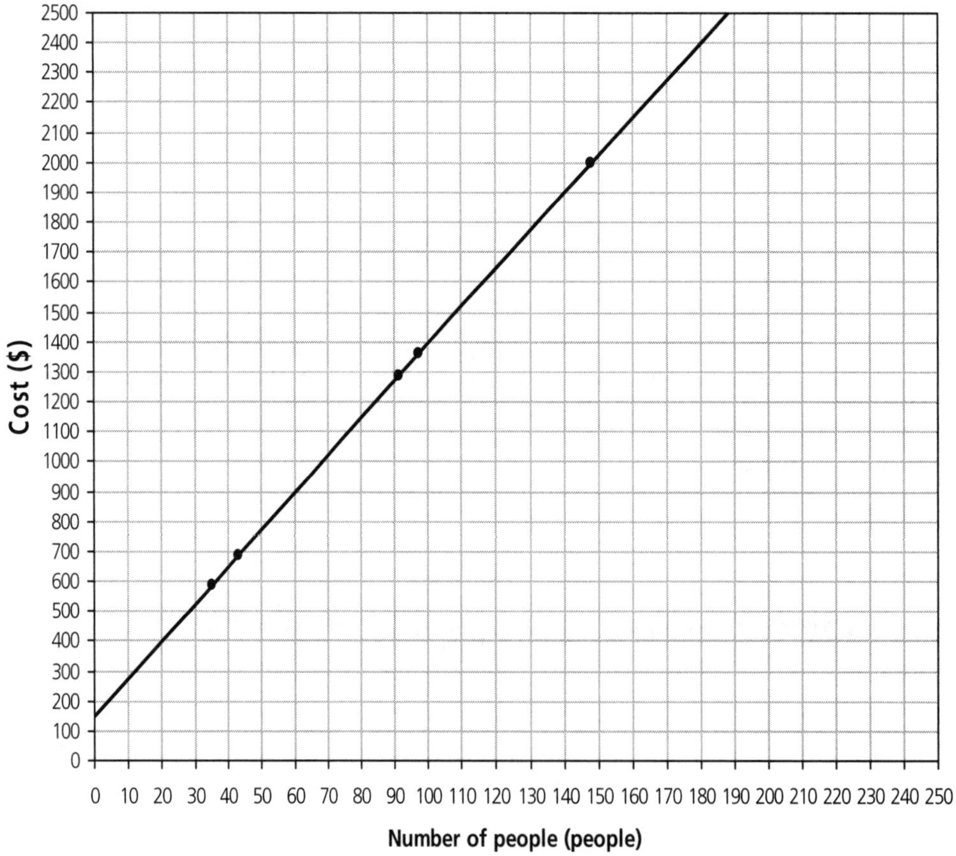

Assignment for Move a Sand Pile

Name: ____________________________

You are riding in a hot-air balloon that is 525 feet in the air. It is descending at the rate of $10\frac{1}{2}$ feet per minute.

1. How many feet high will it be in 10 minutes?

   In 10 minutes the balloon will be at a height of 420 feet.

   525 – 10.5(10)

2. How high will it be in $20\frac{1}{2}$ minutes?

   In 20.5 minutes, the balloon will be at a height of 309.75 feet.

   525 – 10.5 (20.5)

3. How high was it 9 minutes and 30 seconds ago?

   Nine and one half minutes ago, the balloon was at a height of 624.75 feet.

   525 – 10.5(-9.5)

4. How high will it be in 34 minutes?

   In 34 minutes, the balloon will be at a height of 168 feet.

   525 – 10.5(34)

5. When will it land on the ground?

   The balloon will land in 50 minutes.

   0 = 525 – 10.5 x

   -525 = -10.5 x

   50 = x

6. Make a table of values, including an algebraic formula. Then construct a graph for this problem situation.

Labels	***Time***	***Height***
Units	***minutes***	***feet***
Formula	***x***	***525 – 10.5 X***
	10	***420***
	20.5	***309.75***
	-9.5	***624.75***
	34	***168***
	50	***0***
	0	***525***
	20	***315***

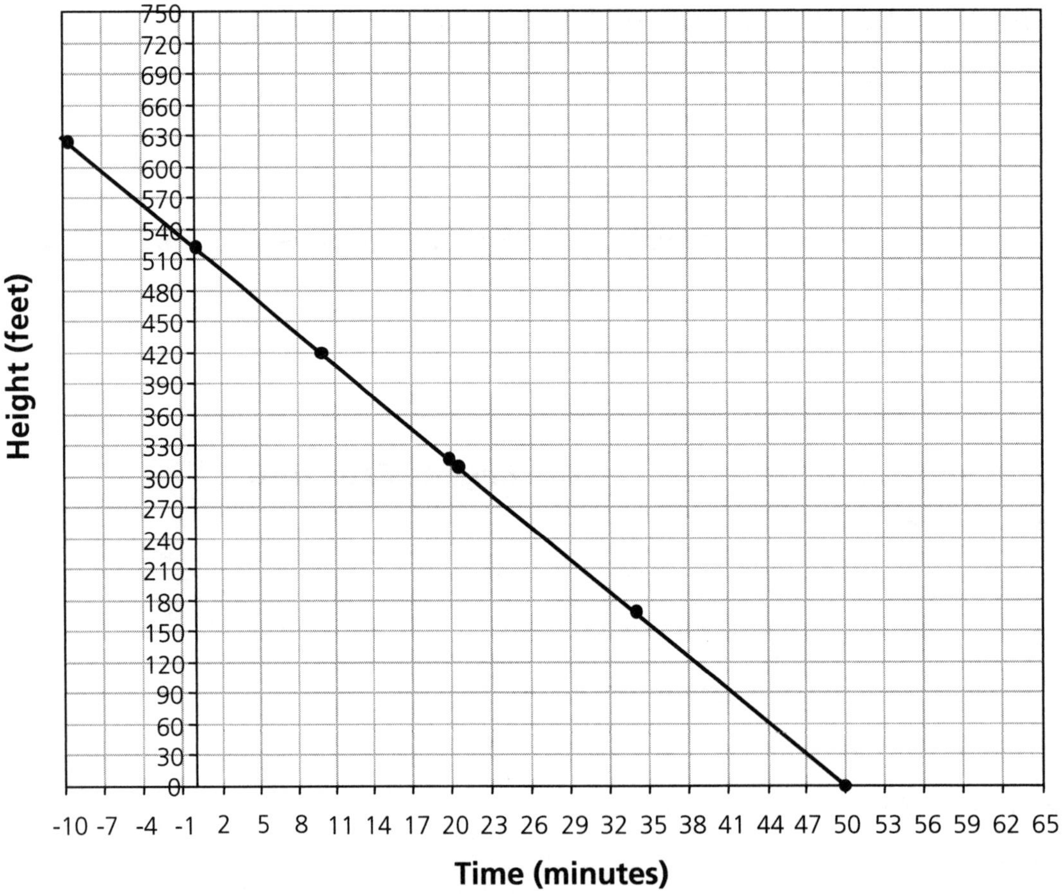

Assignment for Engineering a Highway

Name: ______________________________

Materials to construct birdhouses cost you $50. You plan to sell the birdhouses for $7 each.

Make a table of values to determine profit, including the algebraic formula, and construct a graph for this problem situation.

Table of Values:

Labels	***Birdhouses Sold***	***Profit***
Units	***birdhouses***	***$***
Formula	***b***	***7b – 50***
	0	***-50***
	2	***-36***
	4	***-22***
	6	***-8***
	8	***6***
	10	***20***
	12	***34***

Graph:

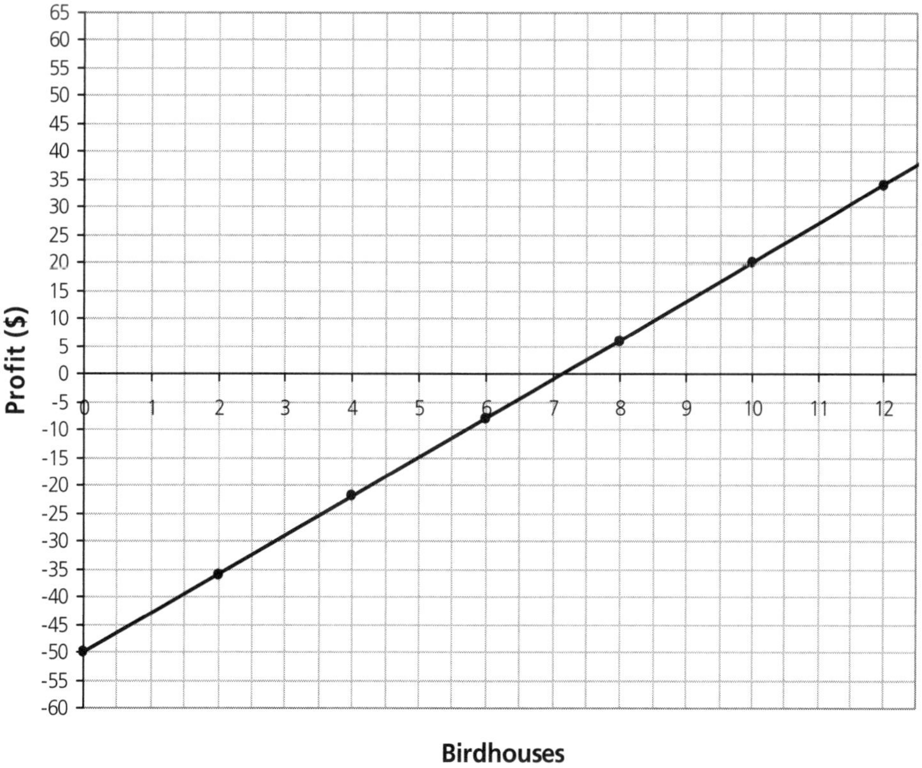

Assignment for Constructing Customized Rain Gutters

Name: ______________________________

Temperatures tend to be similar when observed at the same place and time each year. In a town in Alaska, the average temperature for the fifth of January is -10° Celsius. Each day after that, the average temperature increases at a rate of one-fourth of a degree.

1. What would be the predicted temperature for 30 days after the 5th of January?

   The predicted temperature 30 days after January 5th would be -2.5° Celsius.

   -10 + (0.25)(30)

2. What would be the predicted temperature for 75 days after the 5th of January?

   The predicted temperature would be 8.75° Celsius, 75 days after January 5th.

   -10 + (0.25)(75)

3. What would be the predicted temperature for St. Valentine's Day?
   (Remember that there are 31 days in January.)

   The predicted temperature on Valentine's Day is 0° Celsius.

   -10 + (0.25)(40)

4. What would be the predicted temperature for New Year's Day?

   The predicted temperature on New Year's Day (361 days) would be 80.25° Celsius.

   -10 + (0.25)(361)

5. What would the temperature have been on the previous Christmas Day?

   The temperature on the previous Christmas Day (11 days ago) would have been -12.75° Celsius.

   -10 + (0.25)(-11)

6. Make a table of values, including an algebraic formula. Then construct a graph for this problem situation.

Labels	***Days After January 5th***	***Temperature***
Units	***days***	***degrees Celsius***
Formula	***x***	***-10 + .25x***
	0	***-10***
	30	***-2.5***
	75	***8.75***
	40	***0***
	361	***80.25***
	-11	***-12.75***

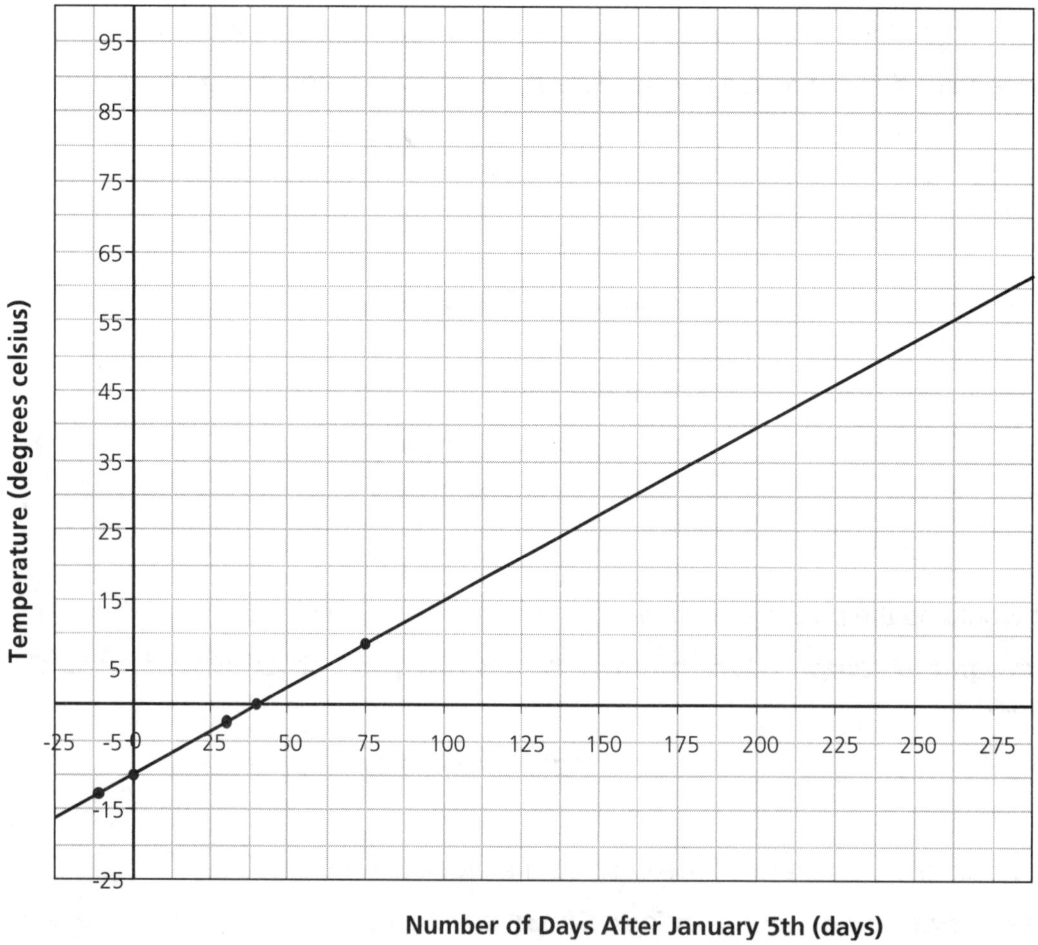

Assignment for Functions

Name: ____________________________

For each problem situation, use function notation to express the relationship between the variables symbolically, and state the domain and range of the function.

1. $8 an Hour Problem

   In this problem situation, you were being paid $8 an hour to assemble furniture at Pat-E-Oh Furniture. You explored the relationship between the number of hours worked during the week and your total earnings for that week.

   a. Designate a letter for the independent variable of the function, and designate a letter for the function name. Use function notation to express the relationship between the variables, using the letters you designated.

   x = time in hours

   f(x) = total earnings

   f(x) = 8x

   b. Based on the number of hours you personally are able to work during any given week, state the domain for this function.

   Answers will vary. In general, domain will be from 0 hours to the maximum number of hours per week.

   c. Based on the domain from part b above, state the range for this function. In other words, what are the least and the most "total weekly earnings" that you personally could achieve in a week?

   The range will be from 0 to 8 times the maximum number of hours from part (b).

2. U.S. Shirts

In this problem situation, you had a job at U.S. Shirts, calculating the cost of various T-shirt orders. For each order, there was a one-time setup charge of $15 and a charge of $8 per shirt. You explored the relationship between the number of shirts ordered and the total cost of the order.

a. Designate a letter for the independent variable of the function, and designate a letter for the function name. Use function notation to express the relationship between the variables, using the letters you designated.

x = shirts ordered

c(x) = cost

c(x) = 8x + 15

b. Suppose U.S. Shirts sets a maximum of 2000 T-shirts per order. State the domain for this function using complete sentences.

The domain is from 0 to 2000 shirts.

c. State the range for this function.

The range is $15 to $16,015.

Assignment for Linear Inequalities

Name: ______________________________

Part 1

To find the ordered pairs on the coordinate plane that satisfy a linear inequality, you first find those that would make the inequality an equation and then decide which of the other two sets of ordered pairs satisfies the inequality.

The procedure that is usually most efficient is to:

- Graph the inequality as if it were an equation
- Find any ordered pair not on this line
- Substitute the ordered pair for the variables in the inequality. If it satisfies the inequality then it is from the solution set. If it does not satisfy the inequality, then the other set must be from the solution set

When the points on the line do *not* satisfy the inequality, you indicate this by drawing the line as a *dashed line* and then *shade in* the area of the coordinate plane that does satisfy the inequality.

Use this procedure to graph the solution set for each of these inequalities.

1. $y > 3x + 4$

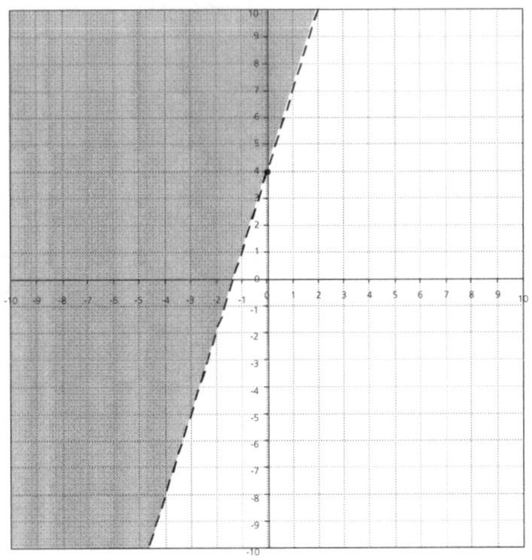

2. $y > 3x$

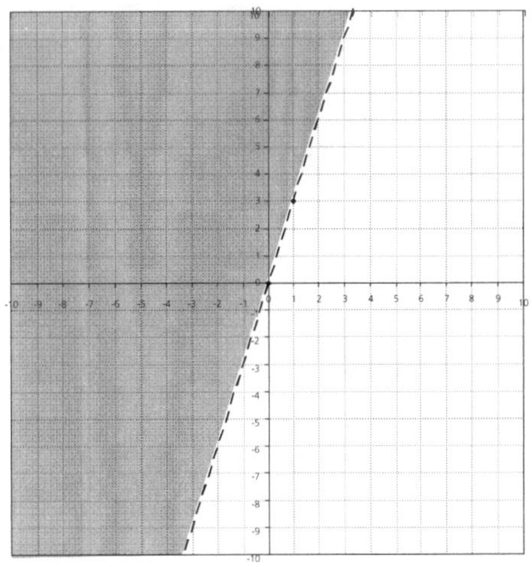

3. $y > x - 4$

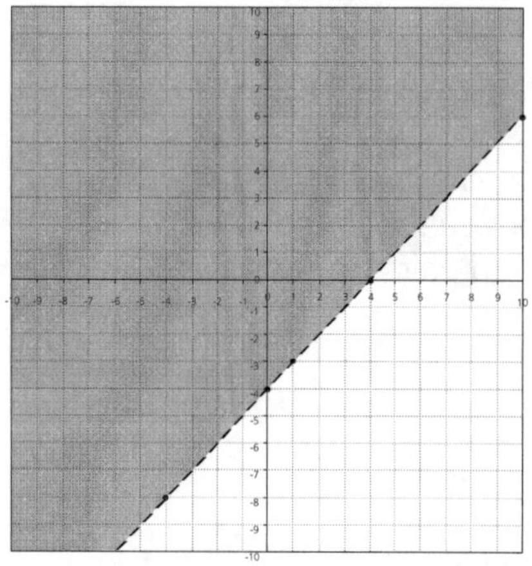

4. $y < 2x - 8$

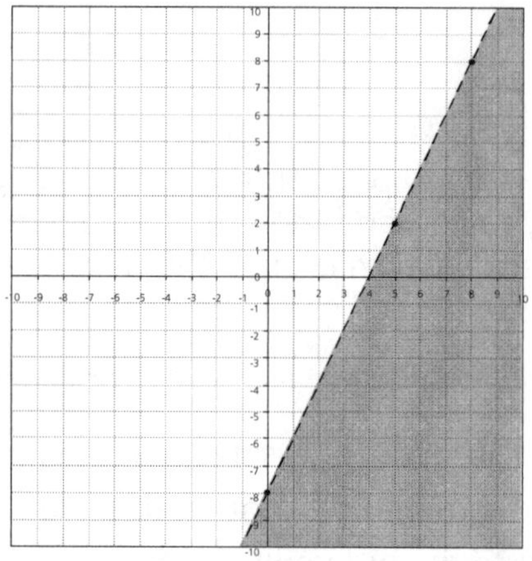

5. $y > 2x - 5$

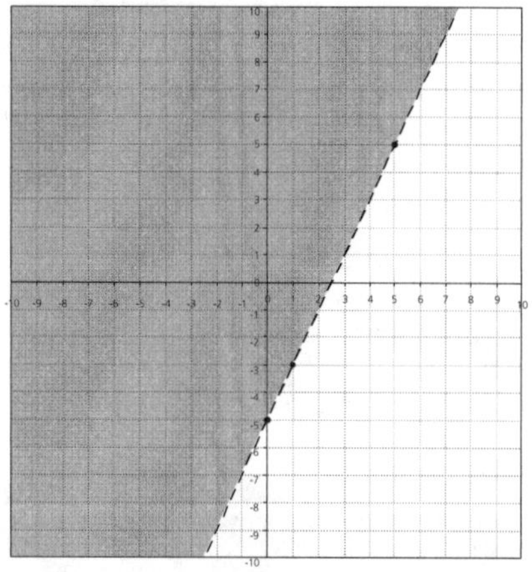

6. $y < x - 2$

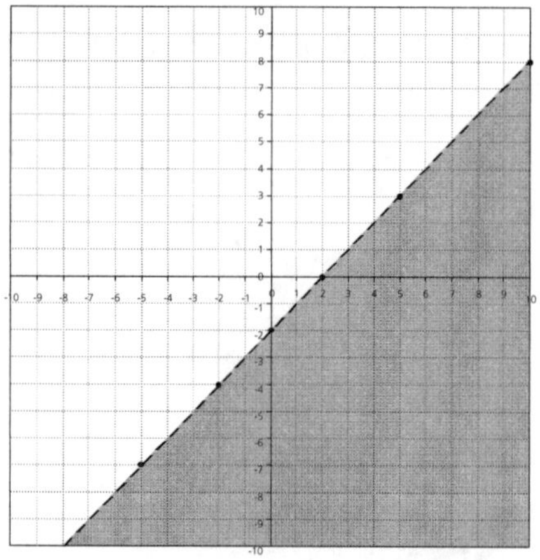

7. $y < 6$

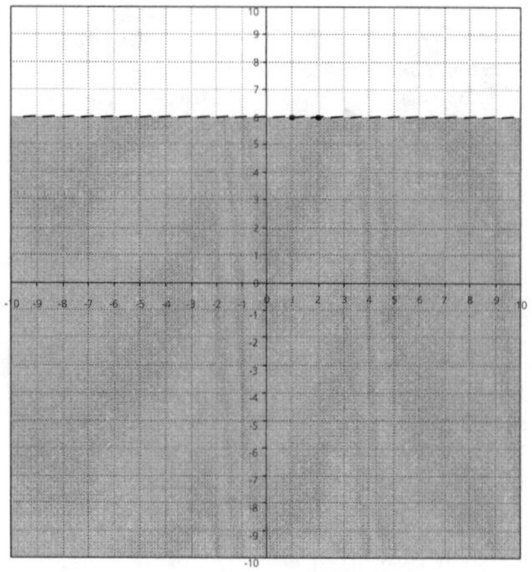

8. $y > -2x$

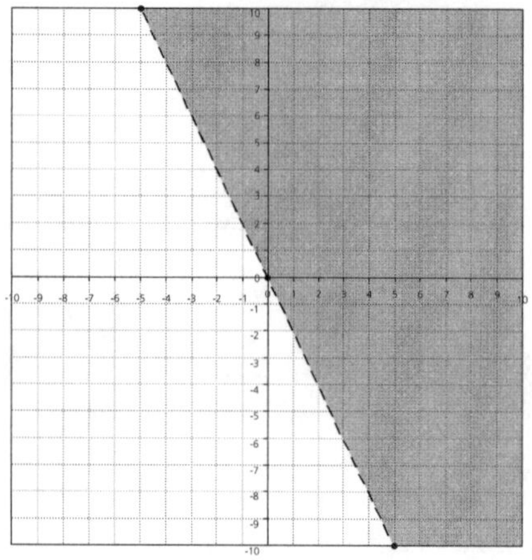

9. $y > 2$

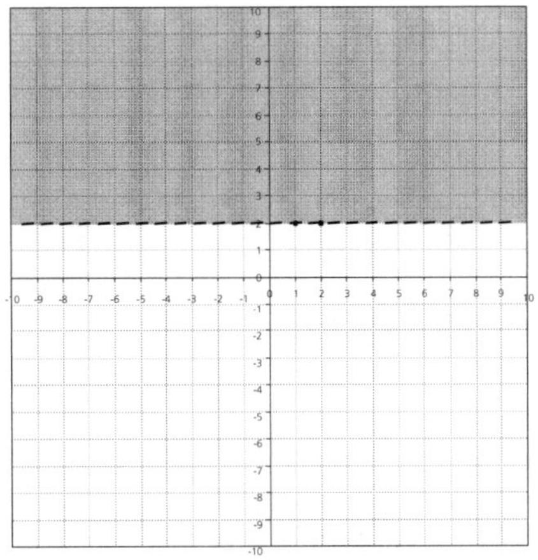

10. $y < -3x + 2$

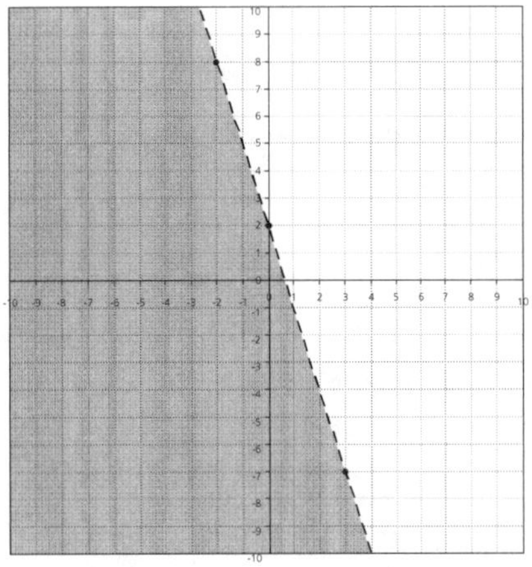

11. $y < -x + 8$

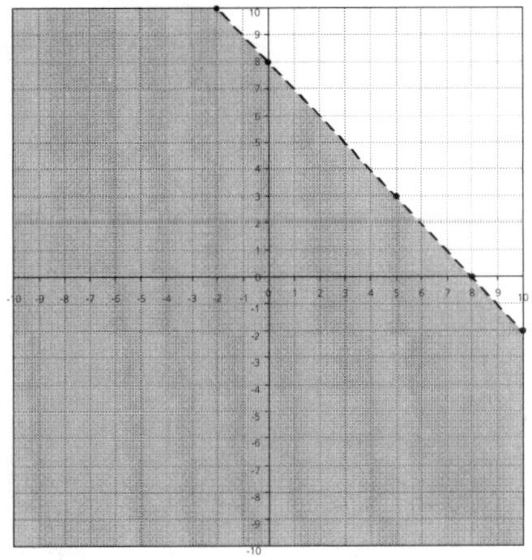

12. $y > -8$

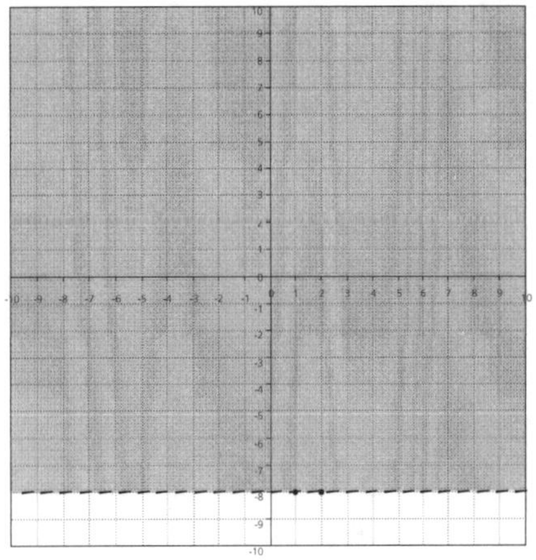

Part 2

Now you will solve inequalities using the other two inequality symbols: ≤ and ≥.

≤ means ***less than or equal to***

≥ means ***greater than or equal to***

The three symbols, <, >, and ≠, are called *strict inequalities* because equality is not included.

Since the two new symbols are ≤, ***less than or equal to***, and ≥, ***greater than or equal to***, ordered pairs that make the inequality equal *are* in the solution set. To graph these solution sets, you indicate that the line is included by drawing a *solid* line and then *shading in* the rest of the solution set.

Graph the solution sets for each of the following linear inequalities.

1. $y \geq -x + 6$

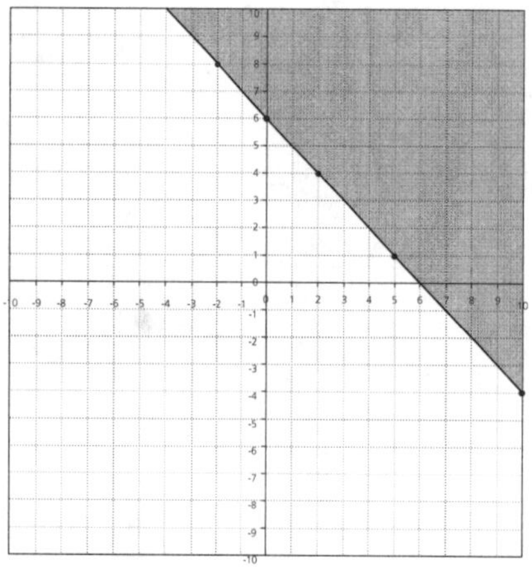

2. $y \leq x + 2$

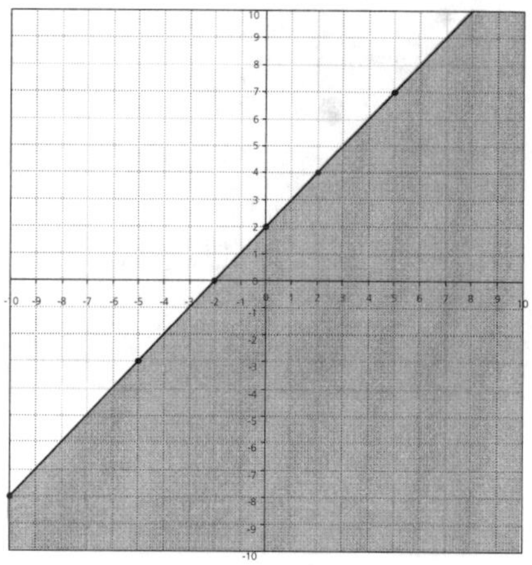

3. $y \geq 3x + 4$

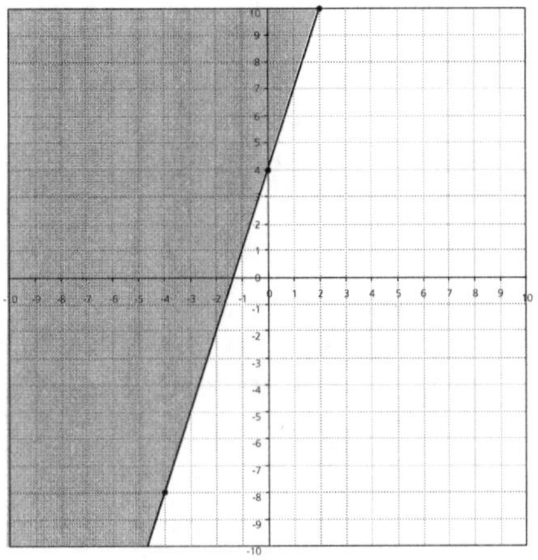

4. $y \leq x - 5$

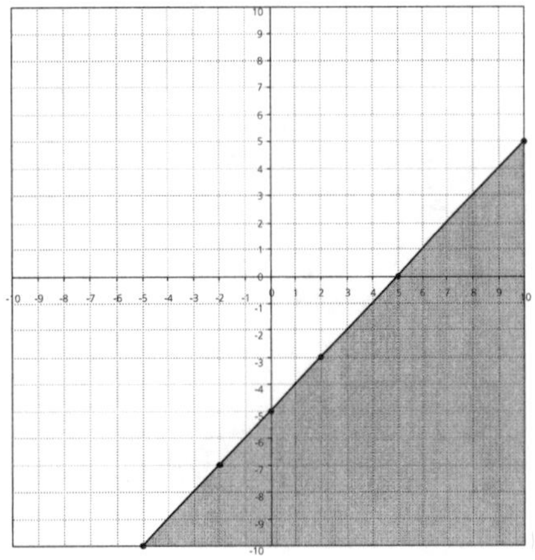

5. $y \geq x + 4$

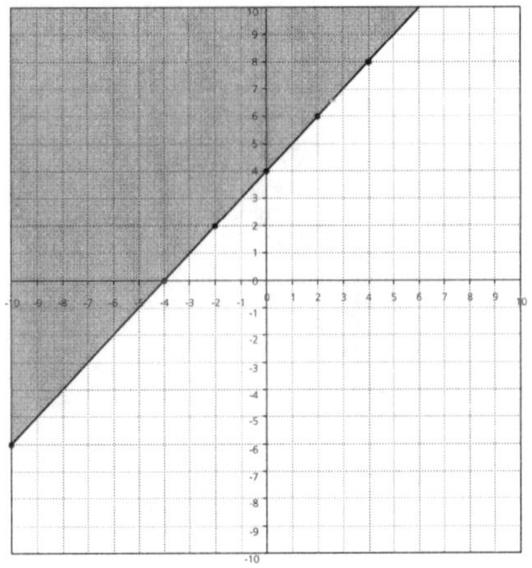

6. $y \leq 3x + 3$

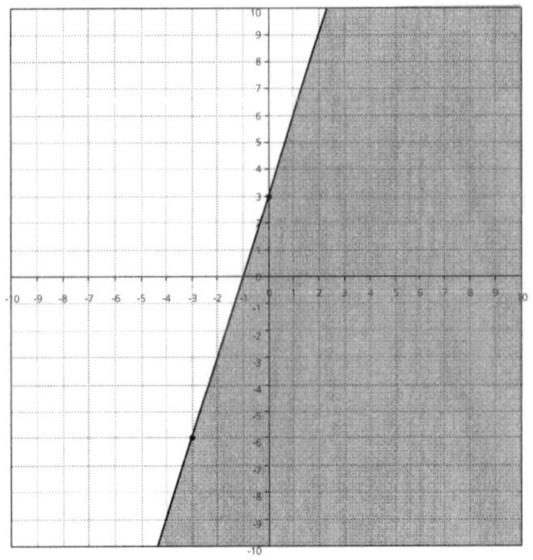

7. $y \leq -x - 2$

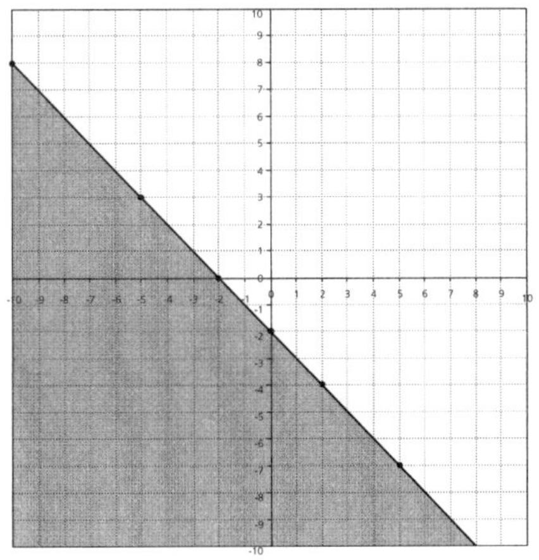

8. $y \geq -x - 2$

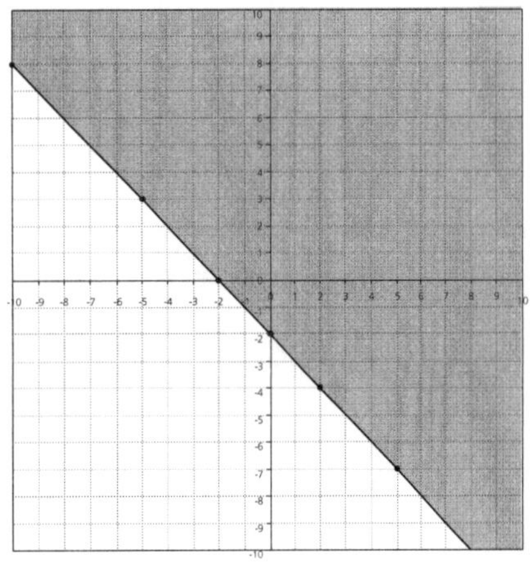

9. $y \leq -2x - 2$

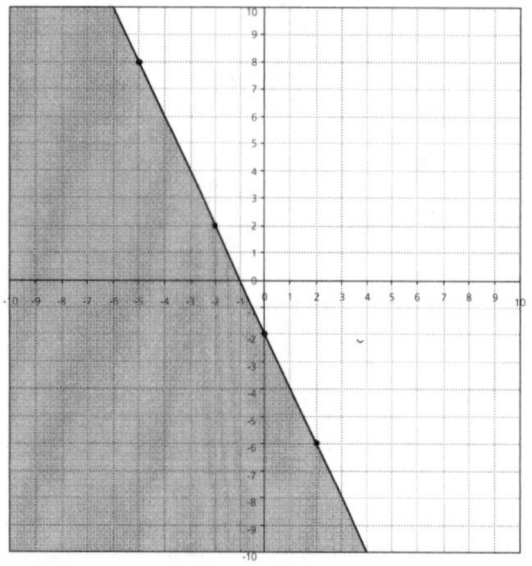

10. $y \geq 4x - 9$

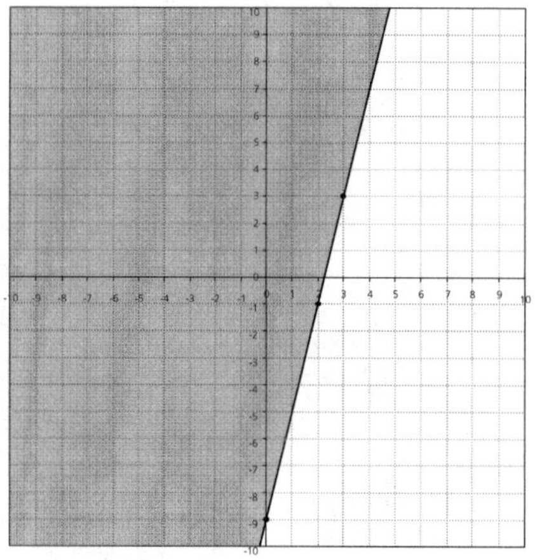

11. $y \leq 3x + 4$

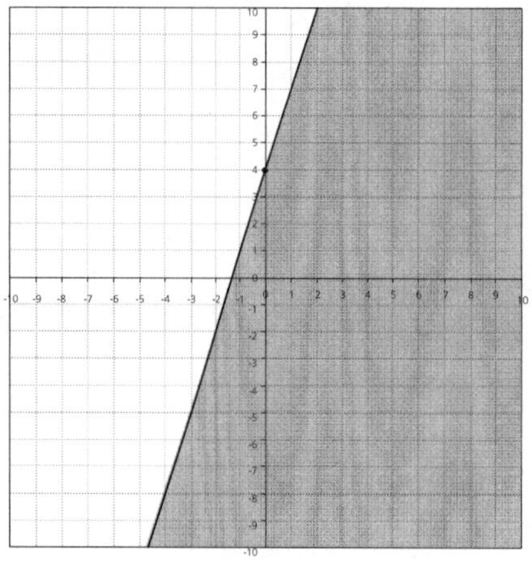

12. $y \geq 5x + 6$

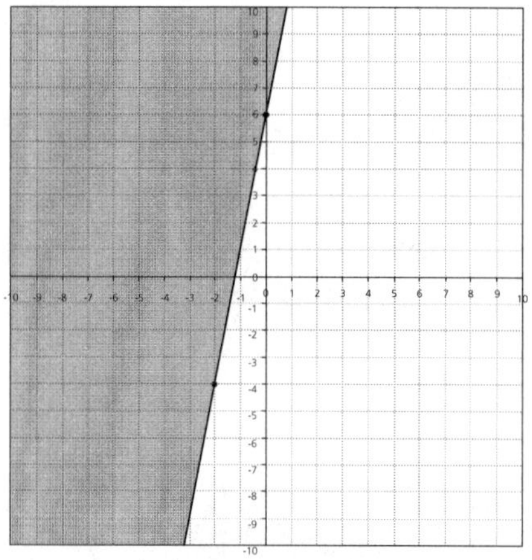

Assignment for Widgets

Name: ______________________________

Widgets cost $7 each, with a shipping charge of $11 per order.

1. Write an algebraic rule for the total cost of an order.

   Total cost = ***$7x + 11$***

2. Find the total cost of an order for:

   a. 12 widgets ***$7(12) + 11 = \$95$***

   b. 257 widgets ***$7(257) + 11 = \$1810$***

3. How many widgets can you order for $88?

   a. Write an equation that states that the cost of the order is $88.

   $$7x + 11 = 88$$

   b. Solve this equation, showing your work.

   $$7x = 77$$

   $$x = 11$$

   c. Write your answer to question 3 in a complete sentence.

   For a cost of $88, 11 widgets can be ordered.

4. How many widgets can you order for $1488?

   a. Write an equation that states that the cost of the order is $1488.

   $$7x + 11 = 1488$$

   b. Solve this equation, showing your work.

   $$7x = 1477$$

   $$x = 211$$

   c. Write your answer to question 4 in a complete sentence.

   For a cost of $1488, 211 widgets can be ordered.

5. How many widgets can you order for $6472?

   a. Write an equation that states that the cost of the order is $6472.

   $$7x + 11 = 6472$$

   b. Solve this equation, showing your work.

   $$7x = 6461$$

   $$x = 923$$

   c. Write your answer to question 5 in a complete sentence.

   Nine-hundred twenty-three widgets can be ordered for a cost of $6472.

Assignments for Dumbbells

Name ______________________________

Dumbbells are sold by the pound. Dumbbells Unlimited sells dumbbells for $0.47 per pound, plus $12 per order for shipping and handling.

1. Write an algebraic rule for the total cost of an order of dumbbells.

   Total cost = ***0.47x + 12***

2. Find the total cost of ordering:

   a. 120 pounds of dumbbells

   0.47(120) + 12

   $68.40

   b. 300 pounds of dumbbells

   0.47(300) + 12

   $153.00

3. How many pounds of dumbbells can you order for $30.80?

   a. Write an equation that states that the cost of the order is $30.80.

   0.47x + 12 = 30.80

   b. Solve this equation, showing your work.

   0.47x + 12 = 18.8

   x = 40

   c. Write your answer to question 3 in a complete sentence.

   Forty pounds of dumbbells can be ordered, for a cost of $30.80.

4. How many pounds of dumbbells can you order for $471.66?

   a. Write an equation that states that the cost of the order is $471.66.

   0.47x + 12 = 471.66

   b. Solve this equation, showing your work.

   0.47x = 459.66

   x = 978

   c. Write your answer to question 4 in a complete sentence.

   Nine-hundred seventy-eight pounds of dumbbells can be ordered, for a cost of $471.66.

5. How many pounds of dumbbells can you order for $48.13?

   a. Write an equation that states that the cost of the order is $48.13.

   $$0.47x + 12 = 48.13$$

   b. Solve this equation, showing your work.

   $$0.47x = 36.13$$

   $$x = 76.87$$

   c. Write your answer to question 5 in a complete sentence.

   Seventy-six pounds of dumbbells can be ordered, for a cost of$48.13.

Assignment for Computer Time and Dumpster

Name: ______________________________

A company has total assets of $675,000. They estimate that these assets are increasing at the rate of $7,500 per week.

1. What will this company's total assets be in ten weeks?

   In 10 weeks, total assets will be $750,000.

2. What were their total assets one year ago?

   One year ago, total assets were $285,000.

3. How many weeks will it take for their assets to double?

   Total assets will double to 1,350,000 in 90 weeks.

4. How many weeks will it take for their assets to reach $750,000?

   Total assets will reach $750,000 in 10 weeks.

5. How long ago were their assets zero?

   Total assets were zero 90 weeks ago.

6. Make a table of values, including an algebraic formula. Then construct a graph for this problem situation.

Labels	***Time***	***Assets***
Units	***weeks***	***dollars***
Formula	***x***	***675,000 + 7,500x***
	0	***675,000***
	10	***750,000***
	-52	***285,000***
	90	***1,350,000***
	10	***750,000***
	-90	***0***

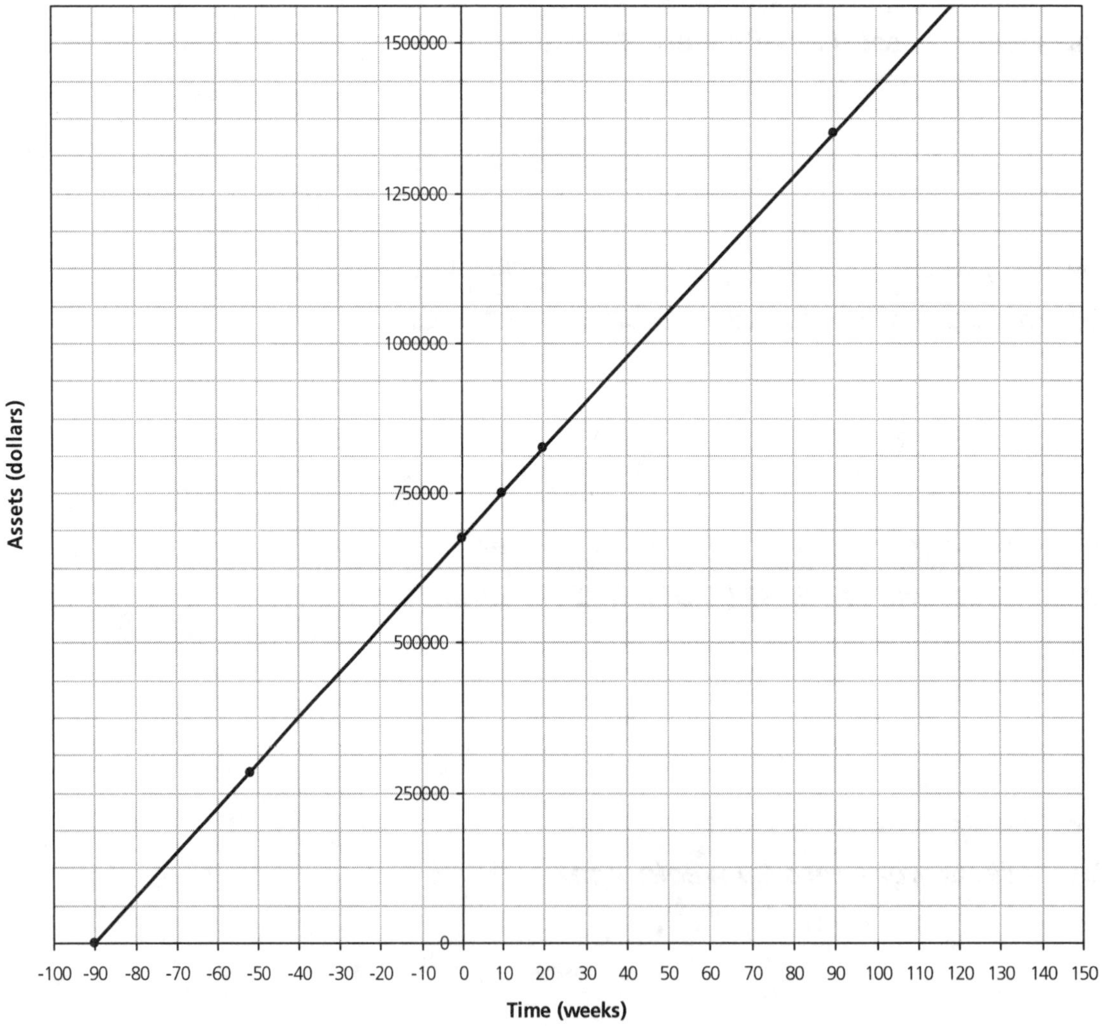

Assignment for Decorating the Math Lab

Name: ______________________________

You have \$20 and are paying out \$2 per day.

1. Find an algebraic equation for this situation.

   Algebraic Equation: ***y = 20 - 2x***

   Define each variable in your equation by writing a short phrase that describes what the variable represents.

   (first variable) ***x*** = ***time (days)***

   (second variable) ***y*** = ***money left (dollars)***

2. Make a table with at least 5 values:

Time (days)	*Money Left (dollars)*
0	*20*
1	*18*
2	*16*
3	*14*
4	*12*

3. Graph the equation for this problem situation:

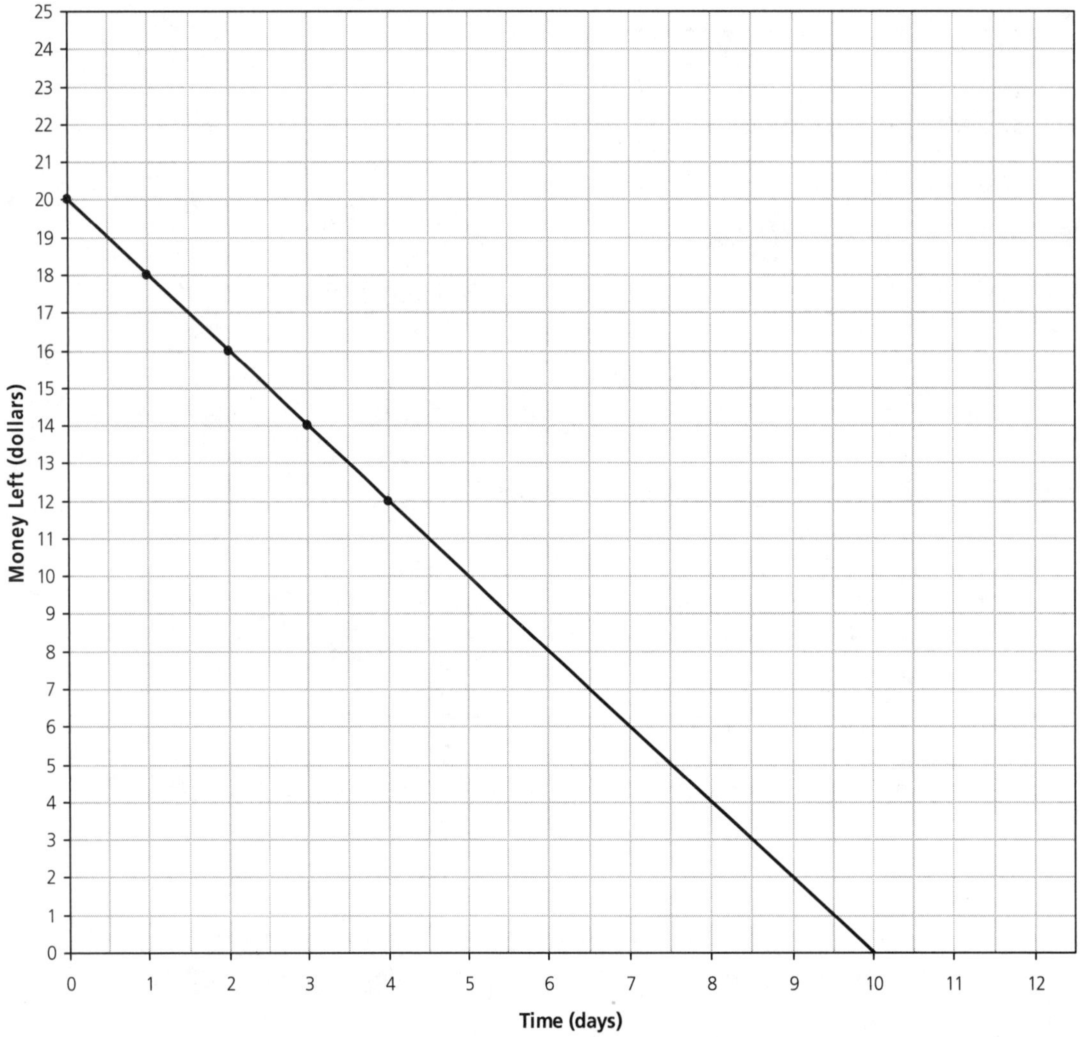

Use your graph, your equation, or your table to answer questions 4 – 8. Write your answers in complete sentences.

4. How would you describe your graph?

   The graph is decreasing.

5. Where does the graph intersect the horizontal axis? Why?

   The graph intersects the horizontal axis at x = 10 because after 10 days all of the money is gone.

6. Where does the graph intersect the vertical axis? Why?

   The graph intersects the vertical axis at y = 20 because you start with $20.

7. What is the value of the second variable when:

   a. the first variable is 7? ***y = 20 - 2(7) = 20 - 14 = 6***

   b. the first variable is -9?

$$y = 20 - 2(-9)$$

$$y = 20 + 18$$

$$y = 38$$

8. What is the value of the first variable when:

   a. the second variable is 16?

$$20 - 2x = 16$$

$$-2x = -4$$

$$x = 2$$

   b. the second variable is -20?

$$20 - 2x = -20$$

$$-2x = -40$$

$$x = 20$$

9. Write an equation that states that the second variable equals 7.5. Solve this equation for the first variable.

$$20 - 2x = 7.5$$

$$-2x = -12.5$$

$$x = 6.25$$

10. Find the value of the second variable when the first variable equals 236.

$$y = 20 - 2(236)$$

$$y = 20 - 472$$

$$y = -452$$

11. Do problems 9 and 10 *make sense* in this problem situation? Why or why not?

    Problem 9 does not make sense because only whole number days are applicable. Problem 10 could make sense if -452 is interpreted as being $452 in debt.

Assignment for Selling Balloons

Name: ____________________________

For each equation, determine whether its graph increases or decreases and find the *rate* of increase or decrease. Then find the intercepts, make a table of values, and sketch a graph of each equation.

Equation: $y = 4x + 4$

Will the graph of this equation increase *(go up)* or decrease *(go down)* from left to right?

The graph will increase from left to right.

What are the intercepts?

x-intercept = ***-1***

y-intercept = ***4***

Whenever x increases by 1, how much does y increase or decrease?

If x increases by 1, y will increase by 4.

Make a table with at least 3 values for x.

x	y
-1	***0***
0	***4***
1	***8***

Sketch a graph of the equation.

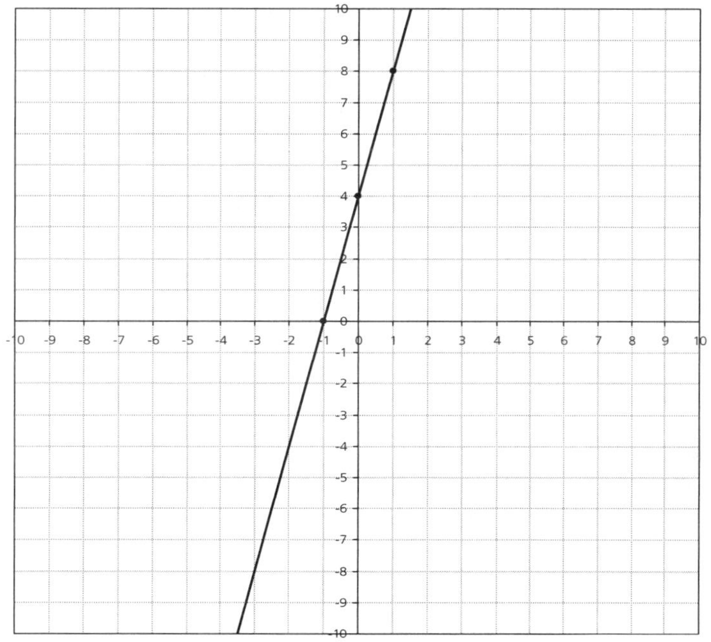

Equation: $y = 4x + 12$

Will the graph of this equation increase *(go up)* or decrease *(go down)* from left to right?

The graph will increase from left to right.

What are the intercepts?

x-intercept = ***-3***

y-intercept = ***12***

Whenever x increases by 1, how much does y increase or decrease?

If x increases by 1, y increases by 4.

Make a table with at least 3 values for x.

x	y
-1	***8***
0	***12***
1	***16***

Sketch a graph of the equation.

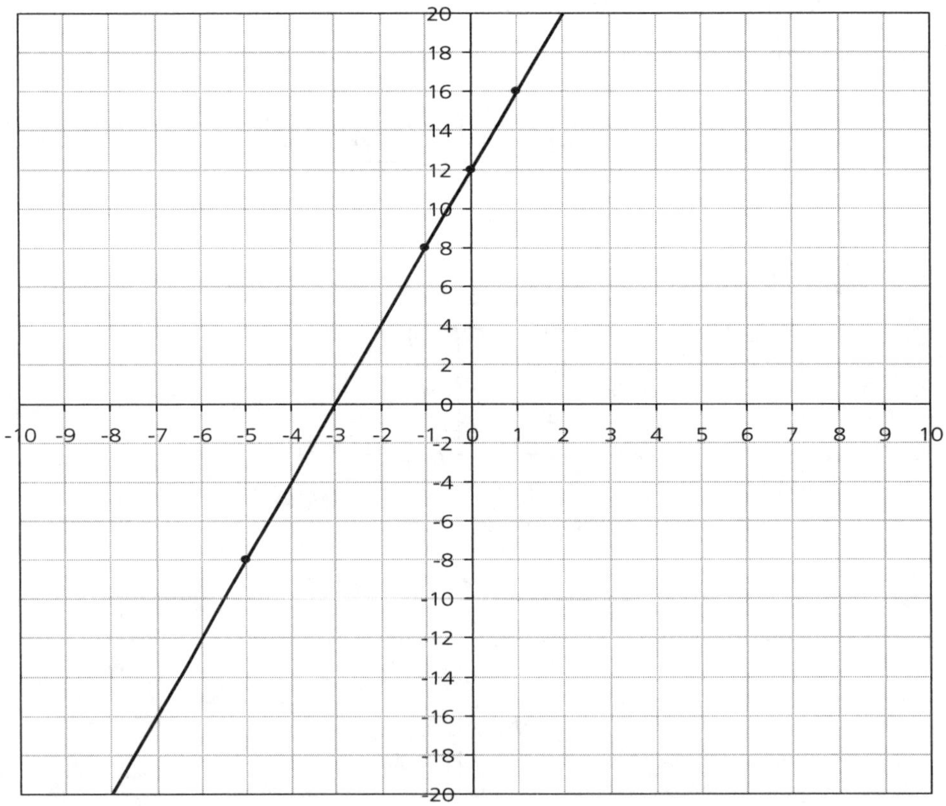

Equation: $y = -x + 5$

Will the graph of this equation increase *(go up)* or decrease *(go down)* from left to right?

The graph will decrease from left to right.

What are the intercepts?

x-intercept = ***5***

y-intercept = ***5***

Whenever x increases by 1, how much does y increase or decrease?

If x increases by 1, y decreases by 1.

Make a table with at least 3 values for x.

x	y
-1	***6***
0	***5***
1	***4***

Sketch a graph of the equation.

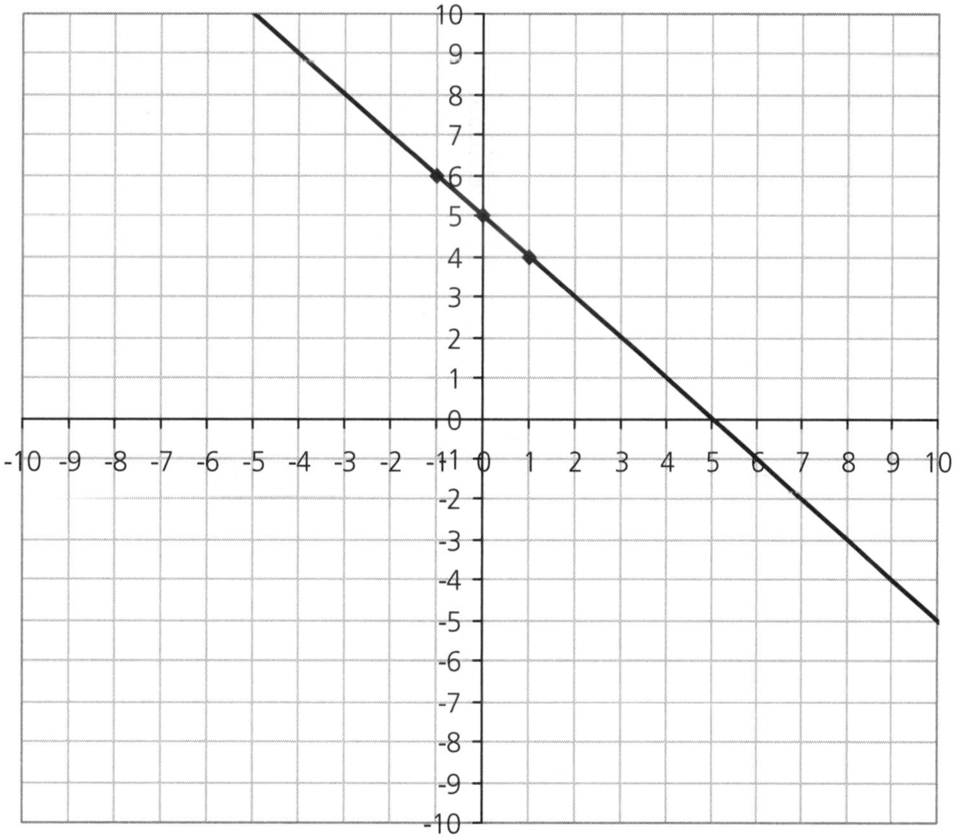

Equation: $y = 9x - 5$

Will the graph of this equation increase *(go up)* or decrease *(go down)* from left to right?

The graph will increase from left to right.

What are the intercepts?

x-intercept = $\boldsymbol{\frac{5}{9}}$

y-intercept = ***-5***

Whenever x increases by 1, how much does y increase or decrease?

If x increases by 1, y increases by 9.

Make a table with at least 3 values for x.

x	y
-1	***-14***
0	***-5***
1	***4***

Sketch a graph of the equation.

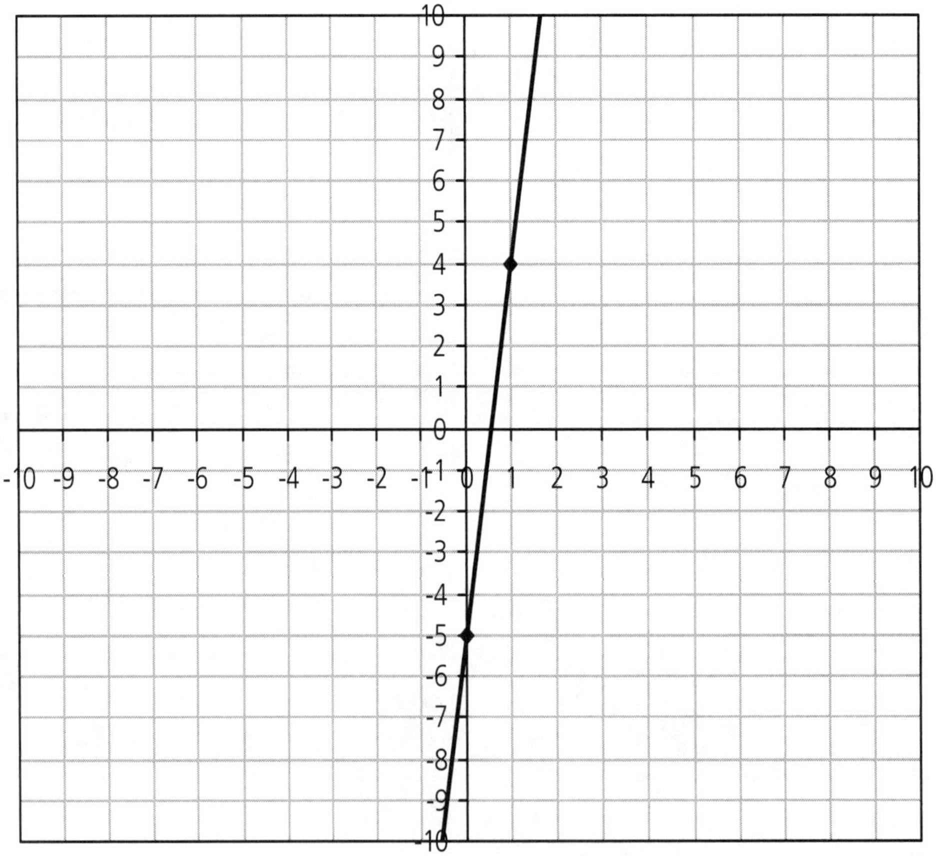

Equation: $y = 10 + 4x$

Will the graph of this equation increase *(go up)* or decrease *(go down)* from left to right?

The graph will increase from left to right.

What are the intercepts?

x-intercept = ***-2.5***

y-intercept = ***10***

Whenever x increases by 1, how much does y increase or decrease?

If x increases by 1, y increases by 4.

Make a table with at least 3 values for x.

x	y
-1	**6**
0	**10**
1	**14**

Sketch a graph of the equation.

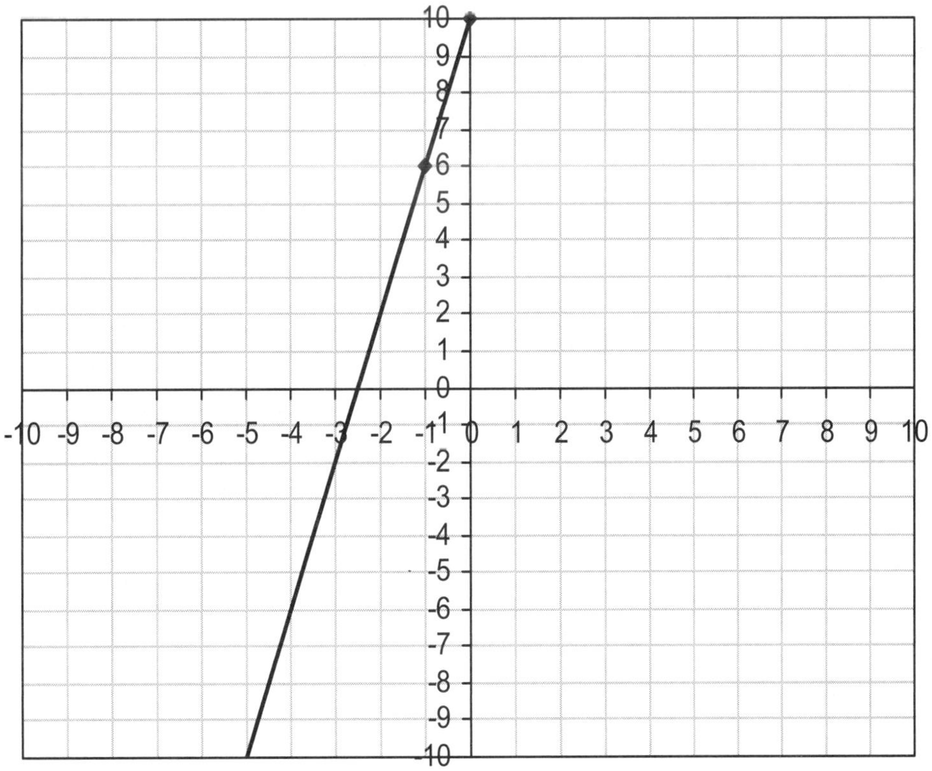

Assignment for Equations, Rates, and Intercepts

Name: ______________________________

For each equation, determine whether its graph increases or decreases and find the *rate* of increase or decrease. Then find the intercepts, make a table of values, and sketch a graph of each equation.

1. $y = -5x + 4$

Will the graph of this equation increase *(go up)* or decrease *(go down)* from left to right?

The graph will decrease from left to right.

What are the intercepts?

x-intercept = $\frac{4}{5}$

y-intercept = ***4***

Whenever x increases by 1, how much does y increase or decrease?

Whenever x increases by 1, y decreases by 5.

Slope: ***-5***

Make a table with at least 3 values for x.

x	y
-1	***9***
0	***4***
1	***-1***

Sketch a graph of the equation.

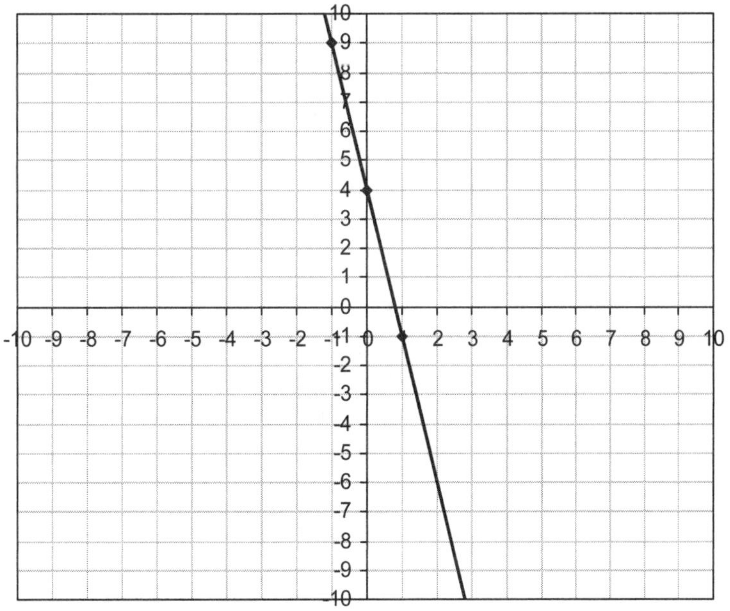

2. $y = 3x - 9$

Will the graph of this equation increase *(go up)* or decrease *(go down)* from left to right?

The graph will increase from left to right.

What are the intercepts?

x-intercept = ***3***

y-intercept = ***-9***

Whenever *x* increases by 1, how much does *y* increase or decrease?

When x increases by 1, y will increase by 3.

Slope: ***3***

Make a table with at least 3 values for *x*.

x	*y*
-1	***-12***
0	***-9***
1	***-6***

Sketch a graph of the equation.

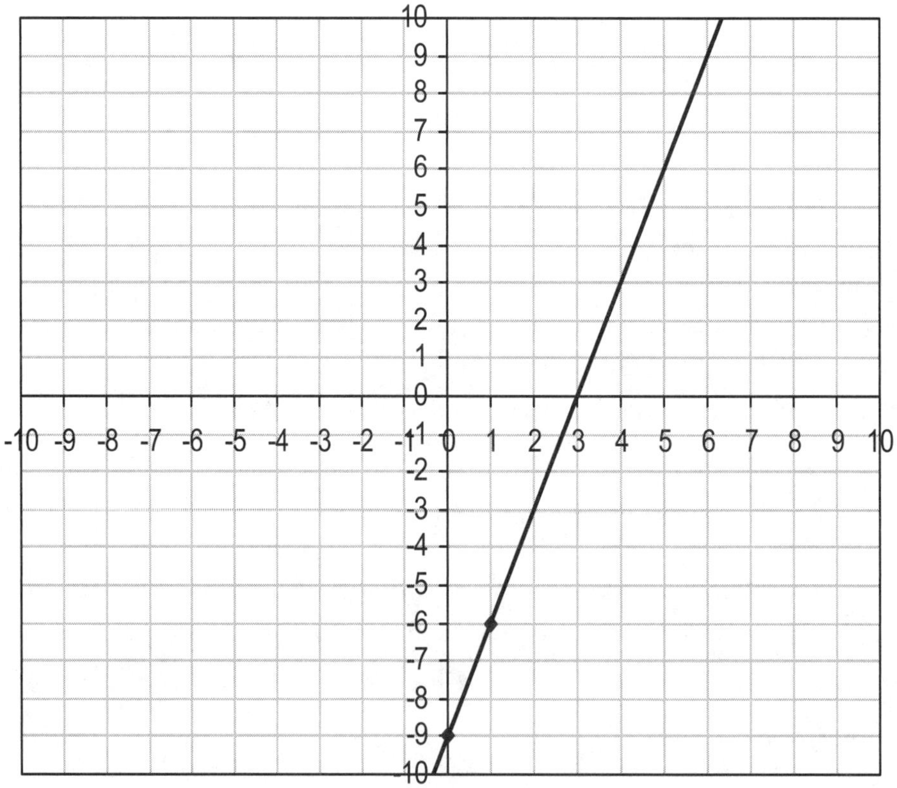

3. $y = 12 + 2x$

Will the graph of this equation increase *(go up)* or decrease *(go down)* from left to right?

The graph will increase from left to right.

What are the intercepts?

x-intercept = ***-6***

y-intercept = ***12***

Whenever x increases by 1, how much does y increase or decrease?

When x increases by 1, y increases by 2.

Slope: ***2***

Make a table with at least 3 values for x.

x	y
-1	***10***
0	***12***
1	***14***

Sketch a graph of the equation.

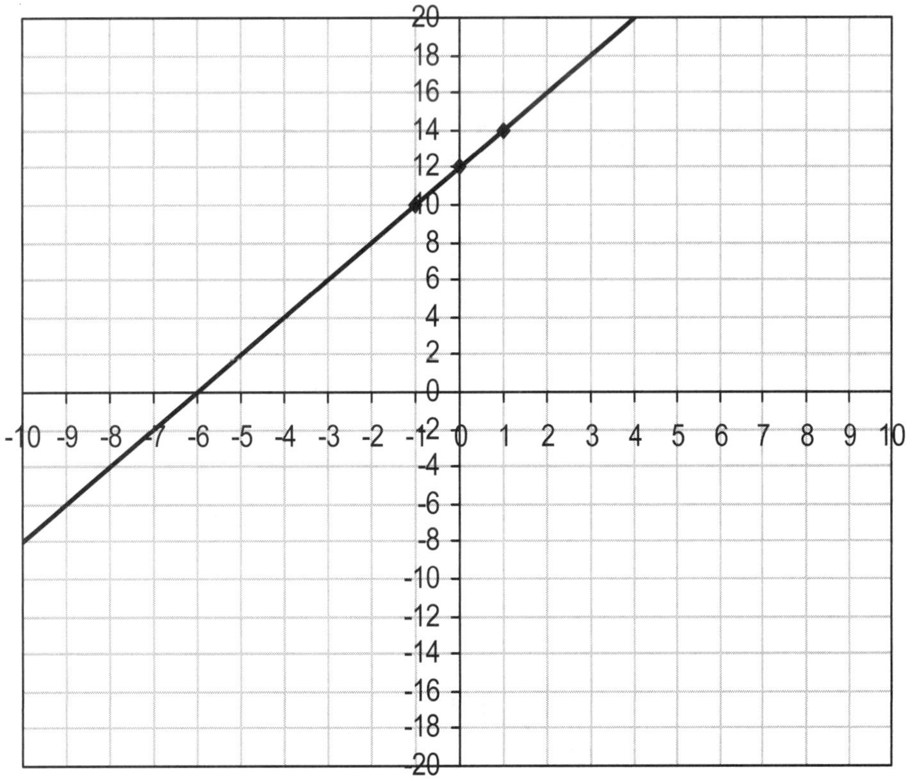

4. $y = 20x - 60$

   Will the graph of this equation increase *(go up)* or decrease *(go down)* from left to right?

   The graph will increase from left to right.

   What are the intercepts?

   x-intercept = ***3***

   y-intercept = -***60***

   Whenever *x* increases by 1, how much does *y* increase or decrease?

   When x increases by 1, y increases by 20.

   Slope: ***20***

   Make a table with at least 3 values for *x*.

x	*y*
-1	***-80***
0	***-60***
1	***-40***

   Sketch a graph of the equation.

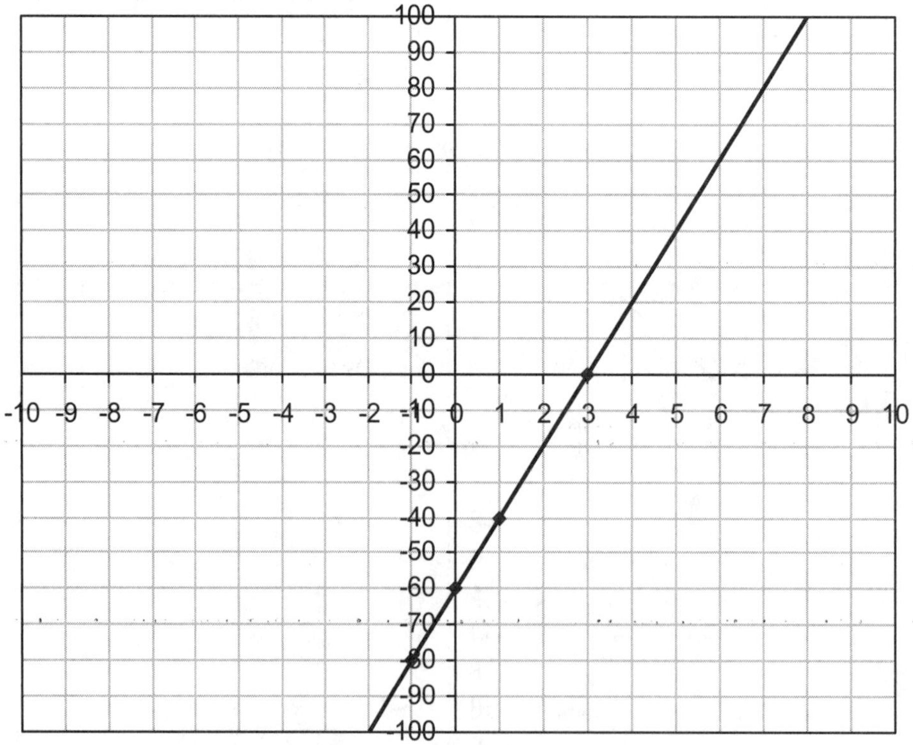

5. $y = 1 - 2x$

Will the graph of this equation increase *(go up)* or decrease *(go down)* from left to right?

The graph will decrease from left to right.

What are the intercepts?

x-intercept = $\frac{1}{2}$

y-intercept = ***1***

Whenever x increases by 1, how much does y increase or decrease?

When x increases by 1, y decreases by 2.

Slope: -***2***

Make a table with at least 3 values for x.

x	y
-1	***3***
0	***1***
1	***-1***

Sketch a graph of the equation.

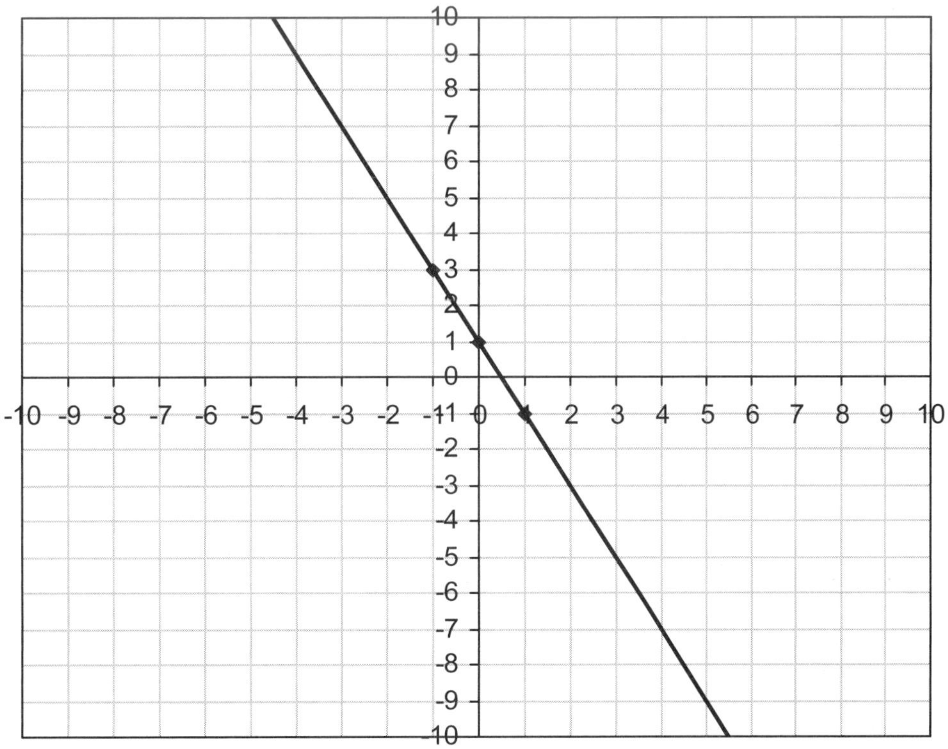

Assignment for Generating and Interpreting Linear Equations

Name: ____

For each equation, find the slope, the *y*-intercept, and the *x*-intercept. Then use the slope-intercept method to graph each line.

1. $y = 2x + 1$

   Slope: **2**

   y-intercept: **1**

   x-intercept: $-\frac{1}{2}$

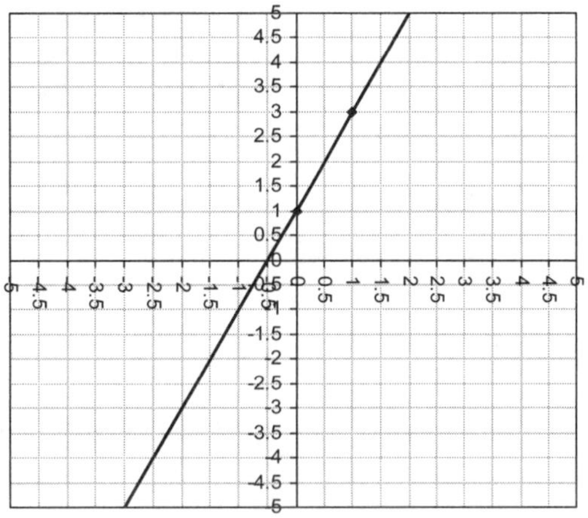

2. $y = 3x$

   Slope: **3**

   y-intercept: **0**

   x-intercept: **0**

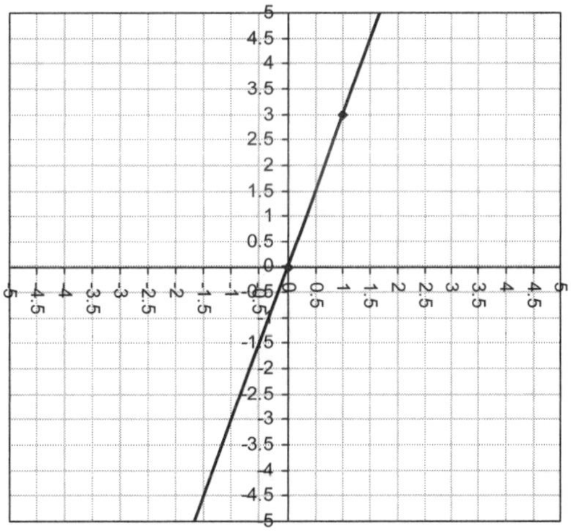

3. $y = x + 4$

   Slope: **1**

   y-intercept: **4**

   x-intercept: **-4**

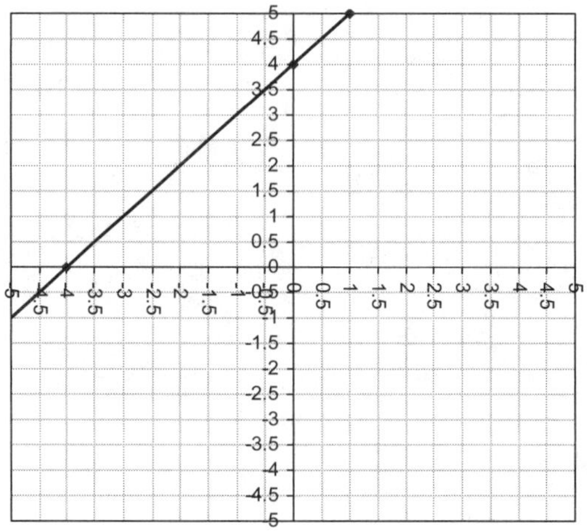

4. $y = 2x - 3$

   Slope: **2**

   y-intercept: **-3**

   x-intercept: $\frac{3}{2}$

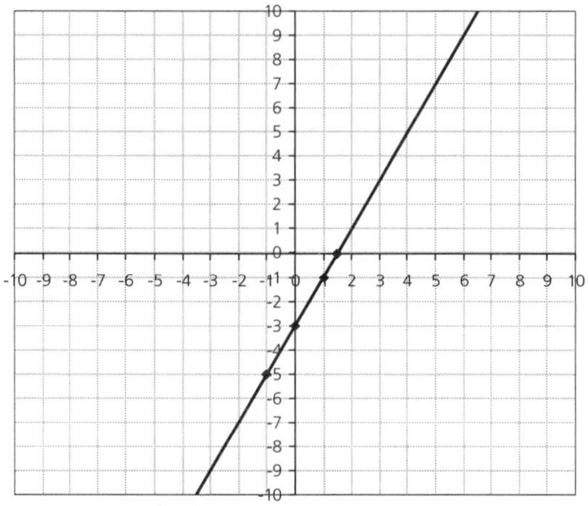

5. $y = \frac{3}{5}x + 2$

   Slope: $\frac{3}{5}$

   y-intercept: **2**

   x-intercept: $-3\frac{1}{3}$

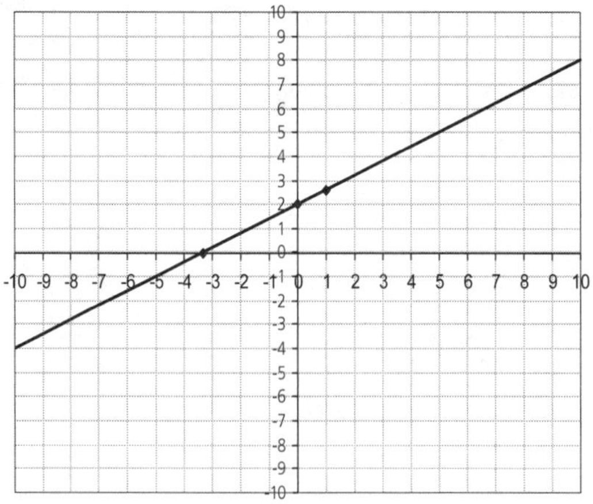

6. $y = 10x + 20$

   Slope: ***10***

   y-intercept: ***20***

   x-intercept: ***-2***

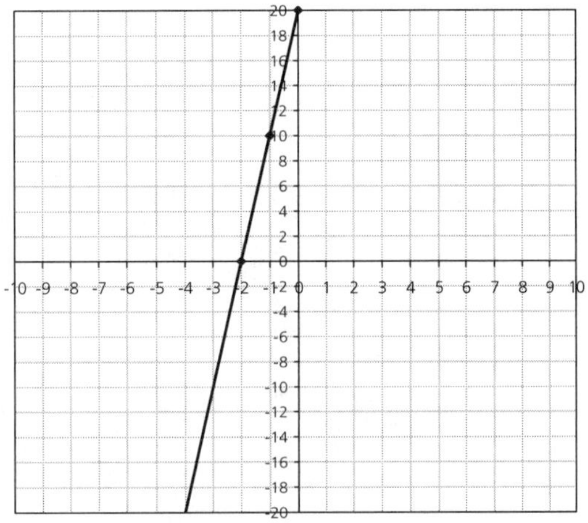

7. $y = 10x + 100$

   Slope: ***10***

   y-intercept: ***100***

   x-intercept: ***-10***

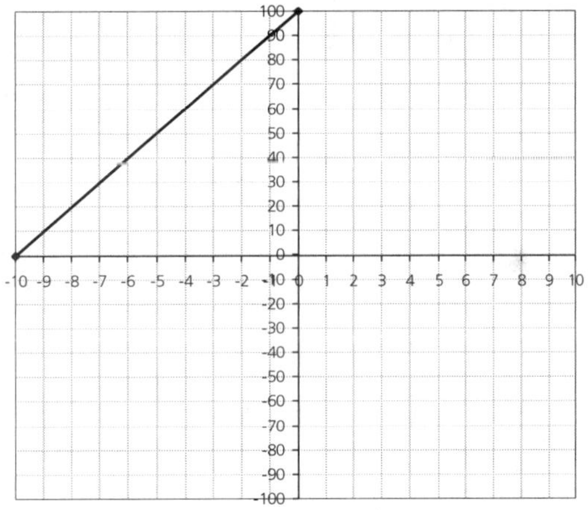

8. $y = 5x - 25$

   Slope: ***5***

   y-intercept: ***-25***

   x-intercept: ***5***

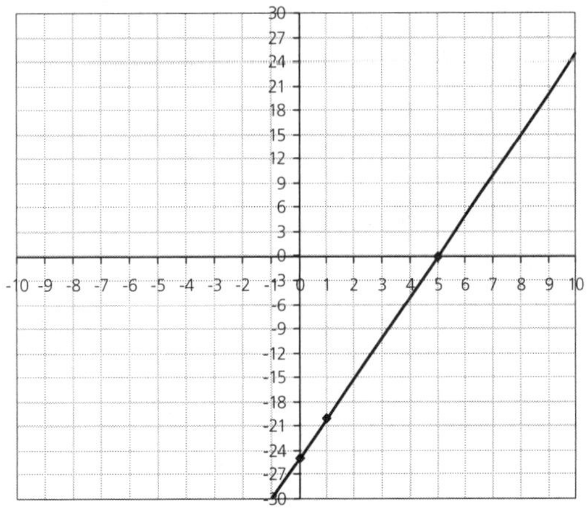

Assignment #1 for Spending Money

Name: ____________________________

In Antarctica, the temperature of the Ross Sea is -10° Celsius at the beginning of October, and goes up about 1° Celsius per week.

1. Find an equation for this situation and write what each variable represents.

   y = -10 + x, where x is time in weeks and y is temperature in degrees Celsius

2. Use complete sentences to answer the following questions.

   a. What is the starting temperature?

   The starting temperature is –10 degrees Celsius.

   b. How many degrees does the temperature go up each week?

   The temperature increases by 1 degree Celsius each week.

   c. In how many weeks will the temperature be 0° Celsius?

   The temperature will be 0 degrees in 10 weeks.

   d. What was the temperature five weeks before the beginning of October?

   Five weeks before the beginning of October, the temperature was –15 C.

   e. What will the temperature be 10 weeks from the beginning of October?

   Ten weeks from the beginning of October, the temperature will be 0 C.

3. Use the equation to find the slope, the *y*-intercept, and the *x*-intercept.

   Slope = ___***1***___ *y*-intercept = ___***-10***___ *x*-intercept = ___***10***___

   Does the graph increase or decrease from left to right? ***The graph increases.***

4. Use complete sentences to answer the following questions.

   a. What does the *y*-intercept represent in this problem situation?

   The y-intercept represents the temperature at the beginning of October.

   b. What does the slope represent in this situation?

   Slope represents how much the temperature increases each week.

   c. What does the *x*-intercept represent in this situation?

   The x-intercept represents when the temperature will be 0 degrees Celsius.

   d. Explain why the graph of this equation increases (or decreases) in this problem situation.

   The graph will increase because the temperature is increasing.

5. Graph your equation in three ways.

   a. First, draw the graph by plotting the y-intercept, the x-intercept, and the point where $x = 1$.

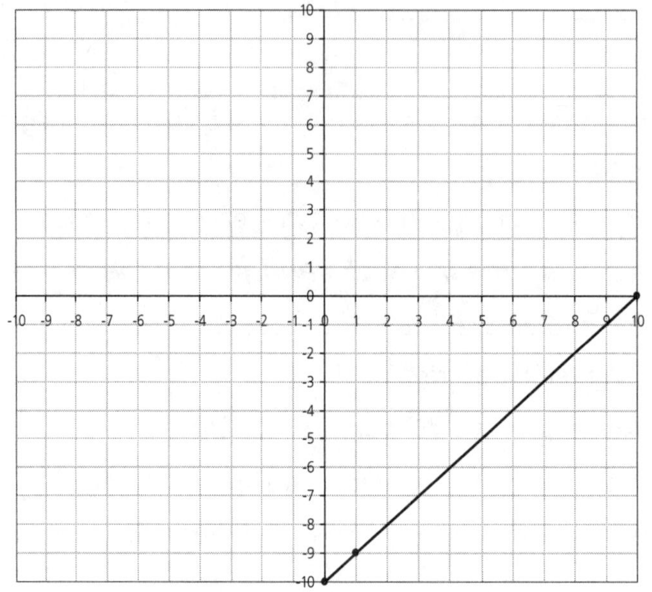

   b. Complete this table. Then use the values in the table to draw the graph on the second set of axes.

x	y
-2	***-12***
2	***-8***
4	***-6***

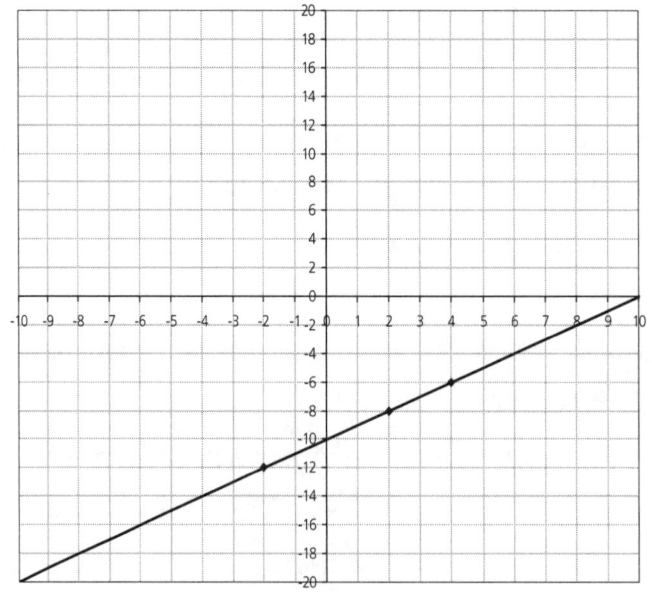

   c. On the third set of axes, draw the graph by plotting the y-intercept and using the slope.

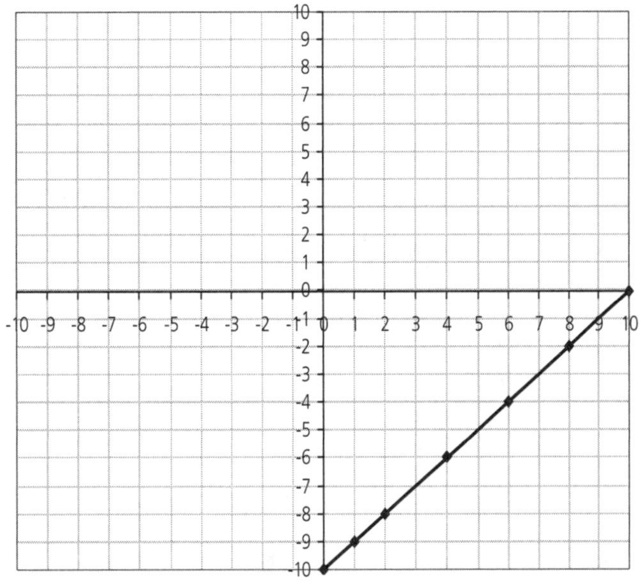

6. Compare these 3 methods of drawing the graph. Which way was easiest for you? Why?

   Answers will vary. Focus on the students' explanation of why the method chosen was easiest.

Assignment #2 for Spending Money

Name: ______________________________

Renting a trailer costs $25 per day, plus $5 per day for insurance and a one-time delivery and pickup fee of $30.

1. Find an equation for this situation that gives the cost for each day you rent the trailer, and write what each variable represents.

   y = 30x + 30, where x is the time in days and y is the cost in dollars

2. Use complete sentences to answer the following questions.

   a. What do you owe when the trailer is delivered?

   You owe $30 when the trailer is delivered.

   b. What does it cost you each day?

   Each day, it costs an additional $30 to rent the trailer.

   c. How many days did you rent the trailer if your cost is $60?

   If you rent the trailer for one day, your cost would be $60.

   d. How many days did you rent the trailer if your cost is $135?

   If your cost is $135, you rented the trailer for 3.5 days.

   e. What is your cost if you rent it for 30 days?

   Your cost to rent the trailer for 30 days is $930.

3. Use the equation to find the slope, the *y*-intercept, and the *x*-intercept.

   Slope = ***30*** *y*-intercept = ***30*** *x*-intercept = -***1***

   Does the graph increase or decrease from left to right? ***The graph increases***.

4. Use complete sentences to answer the following questions.

   a. What does the *y*-intercept represent in this problem situation?

   The y-intercept represents the delivery and pickup fees.

   b. What does the slope represent in this situation?

   The slope represents the total cost per day.

   c. What does the *x*-intercept represent in this situation?

   The x-intercept has no meaning in the context of this problem.

   d. Explain why the graph of your equation increases (or decreases) in this problem situation.

   The graph will increase because the cost to rent the trailer increases as the number of days increases.

5. Graph your equation.

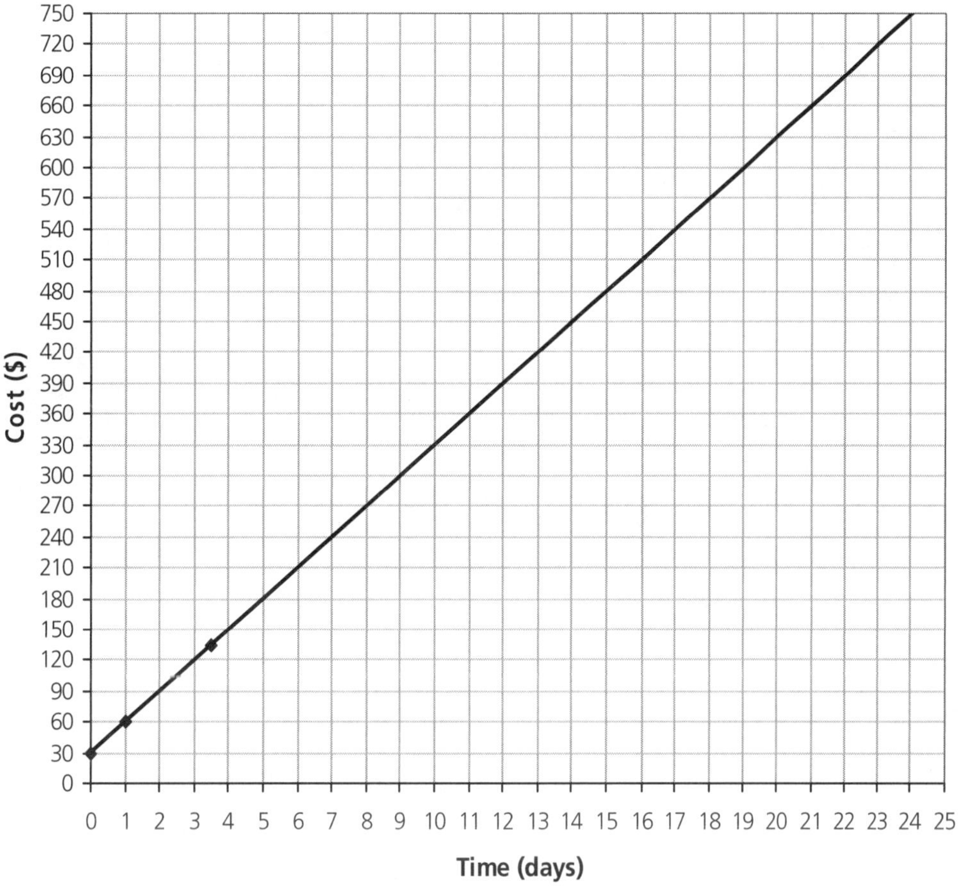

Assignment for Comp-U-Us

Name: ____________________________

Suppose you had sold the first 20 computers for $1,800 each and then sold each additional computer you assembled for $1,800.

Answer the following questions in complete sentences. Show all your work.

1. How much did you receive for the original 20 computers?

   You received $36,000 for the original 20 computers.

2. Define a variable for the number of additional computers you assembled.

   ___***x***___ = Additional computers assembled

3. Use your variable to write a formula for your total income, including what you already received for the original 20 computers.

   1800x+36,000 = Total income

4. Use your formula to find your total income if you assemble and sell:

   a. 20 additional computers ***y = 1800*20+36,000***

   y = $72,000

   b. 50 additional computers ***y= 1800*50+36000***

   y = $126,000

   c. 100 additional computers ***y = 1800*100+36000***

   y = $216,000

   d. 200 additional computers ***y = 1800*200+36000***

   y = $396,000

5. How many additional computers must you assemble to earn a total income of $576,000?

   1800x + 36000 = 576000

   1800x = 540,000

   x = 300 computers

6. Complete this table:

Labels	Additional Computers	Income
Units	computers	dollars
Formula	***x***	***1800x+36,000***
	0	***36,000***
	20	***72,000***
	50	***126,000***
	100	***216,000***
	200	***396,000***
	300	576,000

7. Construct a graph using the number of additional computers and the income for this situation.

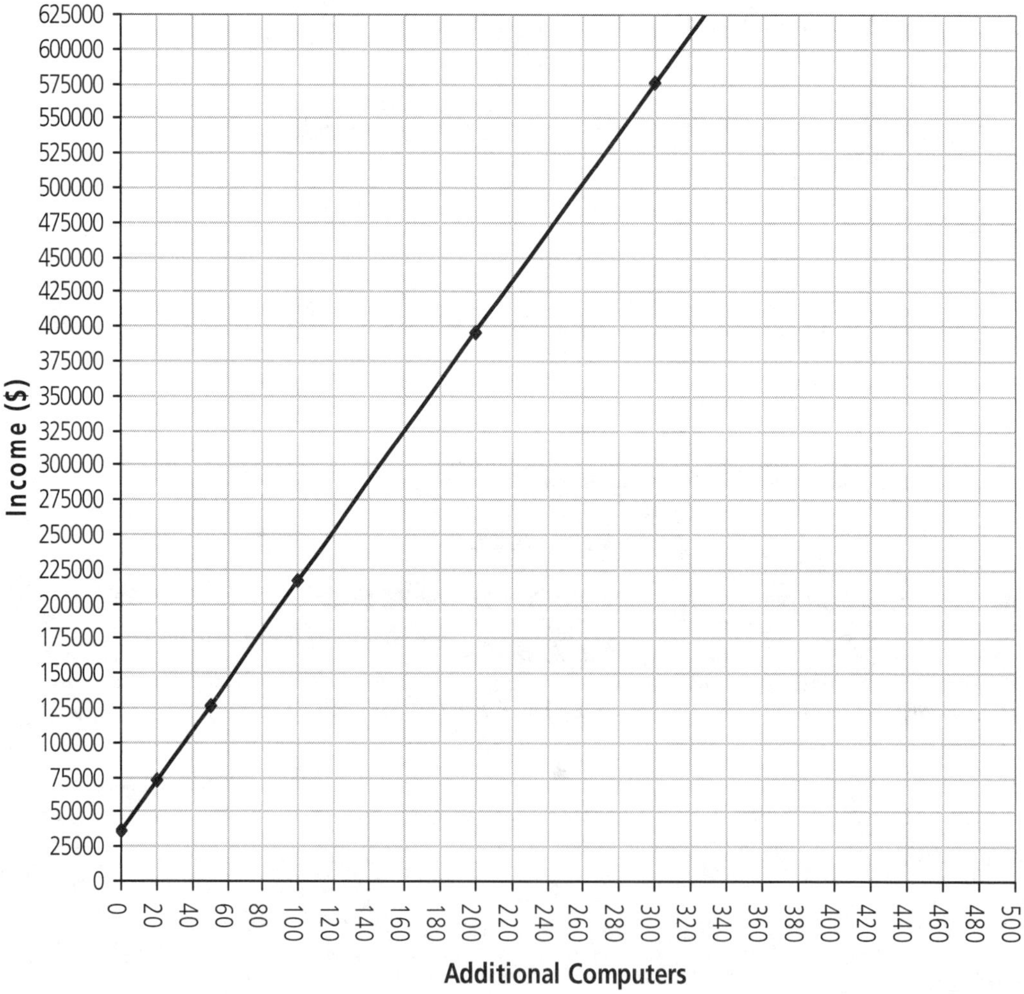

Assignment #1 for Mowing Lawns

Name: ______________________________

Five car salespeople are paid $500, split evenly 5 ways. In addition, each salesperson receives $250 for every car he or she sells.

Complete this table, including the algebraic formula. Draw a line graph illustrating this situation and clearly label the axes.

Labels	***Number of cars sold***	***Money received by each salesperson***
Units	***cars***	***dollars***
Formula	***x***	***250x + 100***
	0	***100***
	10	***2600***
	20	***5100***
	30	***7600***
	40	***10,100***
	50	***12,600***

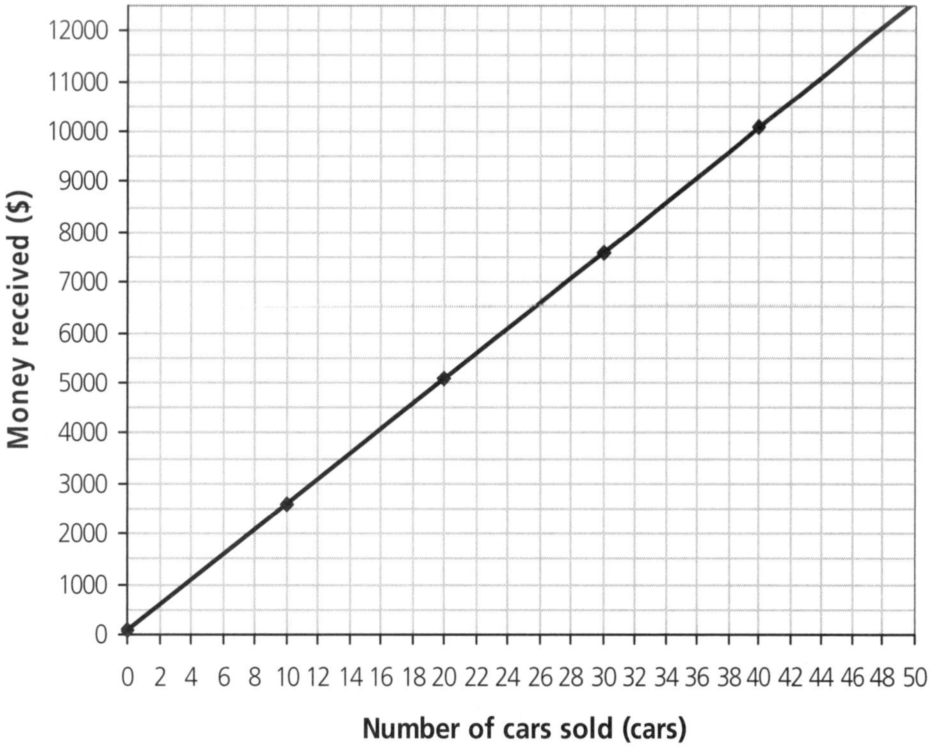

Assignment #2 for Mowing Lawns

Name: ______________________________

You and your two friends have decided to split among yourselves both the cost of your purchases and your income. Your purchases totaled $750.

Use complete sentences to answer the following questions. Show all your work.

1. How much did each partner have to pay for the original purchases?

   Each partner had to pay $250.

2. Define a variable for the amount of income received.

   ______***x***______ = Income received

3. Use your variable to write a formula for each partner's profit (that is, their income minus their share of the purchases).

   $\frac{1}{3}x \quad 250$
   ________________ = Each partner's profit

4. How much will you each receive if your income is:

   a. $750?

   y = (1/3)*750 – 250 = $0

   b. $1,000?

   y = (1/3)*1000 - 250 = $83.33

   c. $2,000?

   y = (1/3)*2000 - 250 =$416.67

   d. $10,000?

   y = (1/3)*10,000 – 250 = $3083.33

5. If each person's profit is $400, what was the total income?

   $\frac{1}{3}x - 250 = 400$

   $\frac{1}{3}x = 650$

   $x = 1950$

   The total income is $1950.

6. Complete this table:

Labels	Total Income	Each Partner's Profit
Units	dollars	dollars
Formula	***x***	$\frac{1}{3}x - 250$
	0	***-250***
	750	***0***
	1,000	***83.33***
	2,000	***416.67***
	10,000	***3083.33***
	1,950	400

7. Construct a graph using the total income and each partner's profit in this situation.

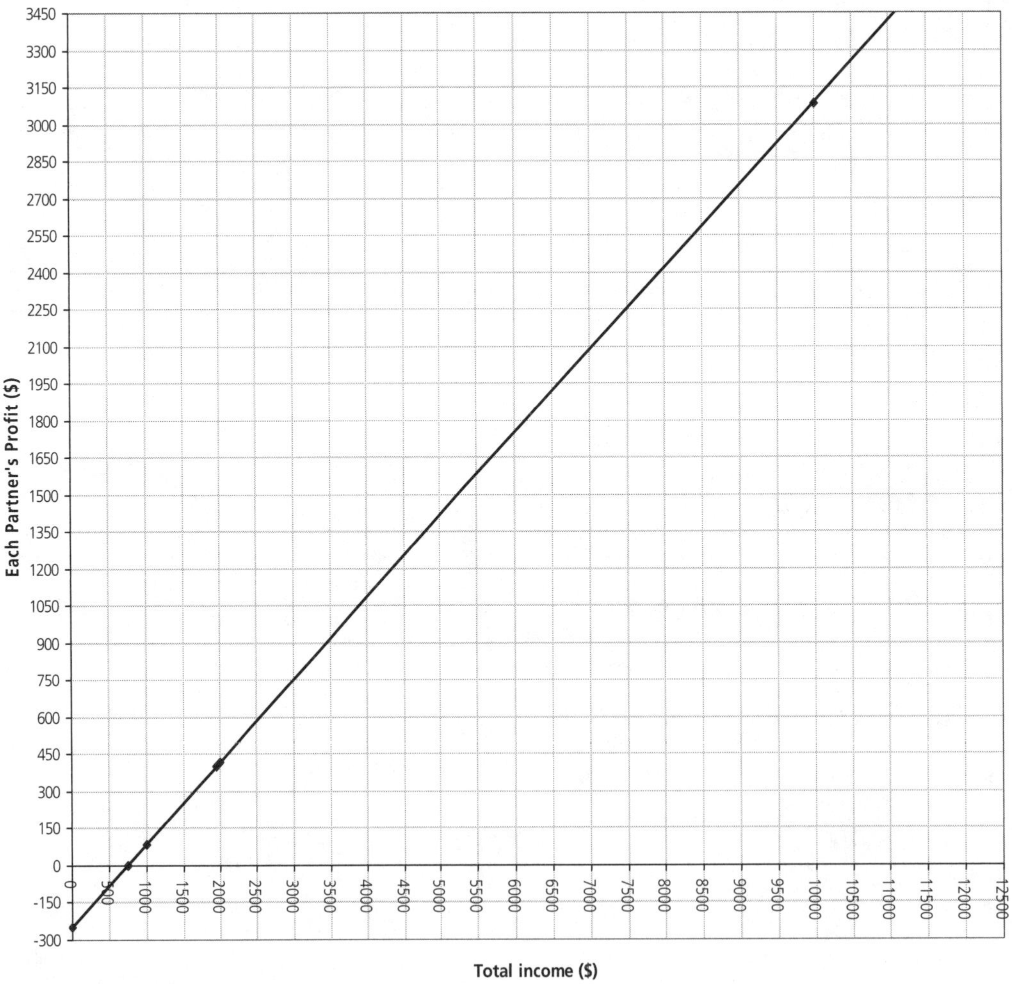

Assignment #1 for Distributive Property

Name: ____________________________

Simplify each expression.

$4(x + 3) =$ ***4x + 12***

$7(x + 3) =$ ***7x + 21***

$2(3x + 4) =$ ***6x + 8***

$(x - 100) \bullet 10 =$ ***10x - 1000***

$6(3x - 7) =$ ***18x - 42***

$8(3x - 4) =$ ***24x - 32***

$5(9 - 2x) =$ ***45 – 10x***

$3(-2x + 5) =$ ***-6x + 15***

$4(-5x + 6) =$ ***-20x + 24***

$6(x + 12) =$ ***6x + 72***

$\frac{36 - 24x}{6} =$ ***6 – 4x***

$\frac{24x - 46}{6} =$ ***4x – (23/3)***

$\frac{56 - 7x}{7} =$ ***8 - x***

$\frac{8x^2 + 40x}{8x} =$ ***x +5***

Factor each expression:

$3x + 6 =$ ***3(x + 2)***

$5x + 80 =$ ***5(x + 16)***

$7x - 28 =$ ***7(x – 4)***

$4x + 18 =$ ***2(2x + 9)***

$10x - 100 =$ ***10(x – 10)***

$28x - 49 =$ ***7(4x – 7)***

$-14x + 7 =$ ***7(-2x + 1)***

$-5 - 15x =$ ***-5(1 + 3x)***

$4x + 7 =$ ***Cannot factor with whole numbers***

$-8x + 6 =$ ***-2(4x – 3) or 2(-4x + 3)***

Assignment #2 for Distributive Property

Name: ____________________________

Use the Distributive Property of Multiplication over Addition (Subtraction) to rewrite each expression.

Example: $5(x + 11) = 5x + 55$

$5(x + 7) =$ **5x + 35**

$7(x + 7) =$ **7x + 49**

$5(5x + 14) =$ **25x + 70**

$(x - 10) \bullet 100 =$ **100x - 1000**

$5(7 - 7x) =$ **35 – 35x**

$7(x + 5) =$ **7x + 35**

$4(x + 6) =$ **4x + 24**

$6(x + 15) =$ **6x + 90**

Use the Distributive Property of Division over Addition (Subtraction) to rewrite each expression.

Example: $\frac{x + 15}{5} = \frac{x}{5} + 3$

$\frac{54 - 36x}{9} =$ **6 – 4x**

$\frac{15x - 23}{3} =$ **5x – (23/3)**

$\frac{16 - 6x}{2} =$ **8 – 3x**

$\frac{9x^3 + 45x^2}{9x^2} =$ **x + 5**

$\frac{15x^2 - 15x}{3x} =$ **5x - 5**

$\frac{400x - 700}{50} =$ **8x - 14**

Factor each expression.

$2x + 6 =$ ***2(x + 3)***　　　$5x + 10 =$ ***5(x + 2)***

$7x - 21 =$ ***7(x – 3)***　　　$4x + 28 =$ ***4 (x + 7)***

$100x - 1000 =$ ***100(x – 10)***　　　$35x - 49 =$ ***7(5x – 7)***

Simplify each expression.

$11x - 6x + 18 - 3x =$ ***2x + 18***

$18 - 3x + 7 =$ ***25 – 3x***

$6(x - 4) + 3(8 - x) =$ ***6x – 24 + 24 – 3x = 3x***

$18 - 6(x - 3) =$ ***18 – 6x + 18 = 36 – 6x***

$-7(8x - 4) + 11 =$ ***-56x + 28 + 11 = 39 – 56x***

$8(3x + 6) - 3(8x + 6) =$ ***24x + 48 – 24x – 18 = 30***

Assignment for Deck Problem

Name ______________________________

Find the area of your hand by tracing it on this grid. How does the area of your hand compare to the area of the hand of the person sitting next to you? Can you predict who in your class has the hand with the least area? Who has the greatest area? Recall the strategies you used in the previous problem so that you do not have to count each individual square to find the area of your hand.

Answers will vary. The student with the largest hand will have the largest area. The person with the smallest hand will have the smallest area.

Assignment for Area of Rectangle

Name: ______________________________

1. These rectangular boxes need to be wrapped. To see how much wrapping paper you need, find the total surface area of each box. Figures are not drawn to scale. (Each box has six sides.)

a.

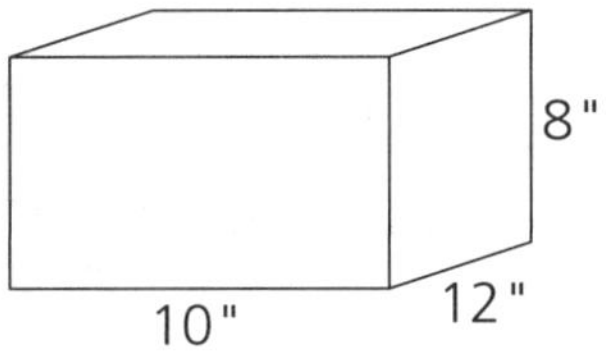

front = 10(8) = 80

top = 10(12) = 120

side = 8(12)=96

surface area = 2(80 + 120 + 96) = 592 in^2

b.

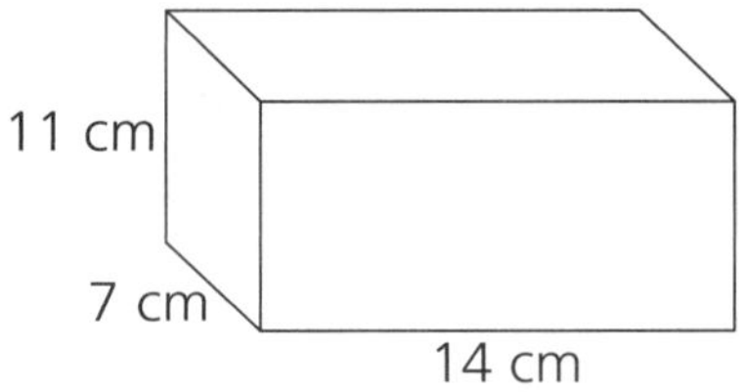

front = 11(14) = 154

top = 14(7) = 98

side = 11(7) = 77

surface area = 2(154 + 98 + 77) = 658 cm^2

c.

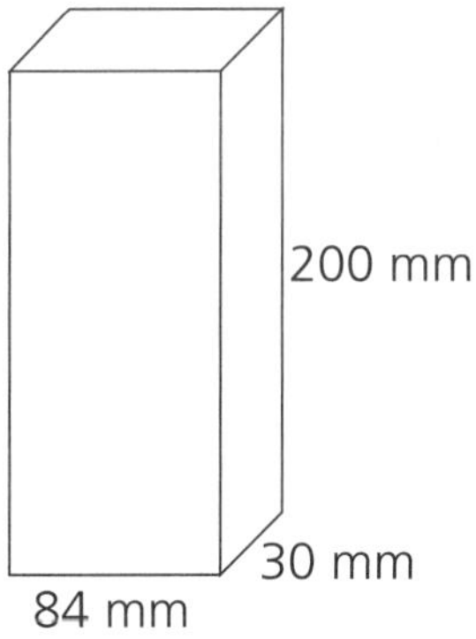

front = 84(200) = 16,800

top = 84(30) = 2520

side = 30(200) = 6000

surface area = 2(16,800 + 2520 + 6000) = 50,640 mm^2

2. In the figure, a side of the inner square measures 35 inches. A side of the outer square measures 40 inches. Find the area of the shaded region.

inner square = 35(35) = 1225

outer square = 40(40) = 1600

shaded region = 1600 – 1225 = 375 in^2

3. In the figure, the outer rectangle is 4 feet by 10 feet. The inner rectangle is 3 feet by 9 feet. Find the area of the shaded region.

inner rectangle = 3(9) = 27

outer rectangle = 4(10) = 40

shaded = 40 – 27 = 13 ft²

4. Measure the figure below with a centimeter ruler to find the area of the shaded region.

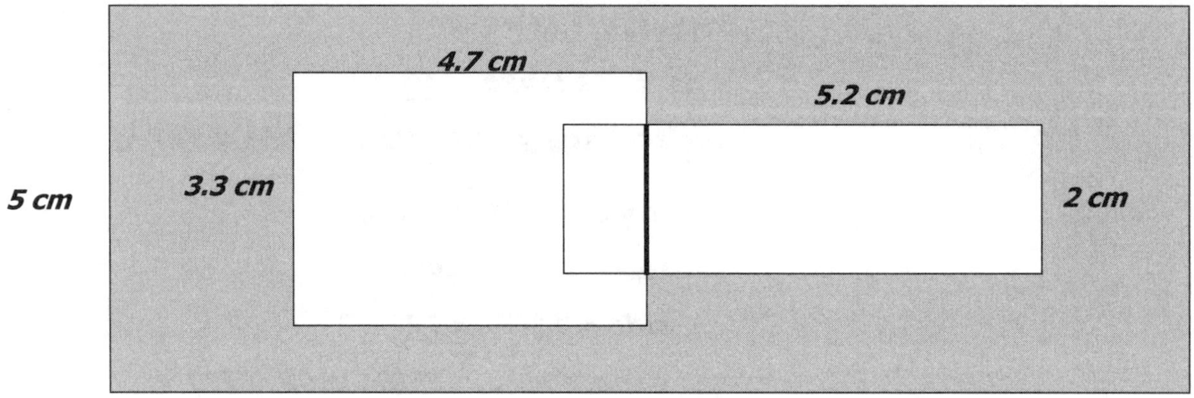

large figure = 5(14.3) = 71.5

figure (left) = 4.7(3.3) = 15.51

figure (right) = 5.2(2) = 10.4

shaded region = 71.5 – 15.51 – 10.4

= 45.59 cm²

Answers may vary based upon variation in student measurement.

5. The area of a rectangular field is 340 square miles. If the length of the field is 20 miles, find the width of the field. Write a sentence explaining how you arrived at your answer.

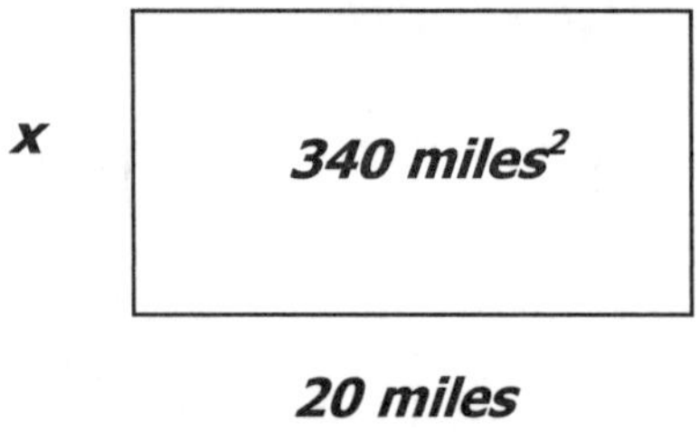

20x = 340
x = 17

6. The area of a square garden is 676 square feet. Find the perimeter of the garden. Write a sentence explaining how you arrived at your answer.

   S^2 = 767

   S = 26

   perimeter = 4(26) = 104

   The perimeter of the garden is 104 feet.

7. You just bought a roll of wrapping paper measuring 25 square feet. The width of the roll is 1 foot. What are the dimensions of the largest rectangular box you can wrap with this roll of paper? Write a sentence explaining how you arrived at your answer

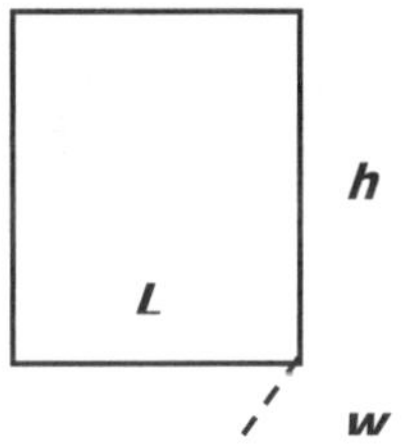

Answers will vary based on each student's interpretation of what qualifies as "the largest rectangular box." Using the diagram, 2h + 2w cannot exceed 25 feet and 2l +2w cannot exceed 2 feet. One possible box would have a square base of side length 6 inches and a height of 12 feet.

8. You are designing a red and black checkerboard banner. The area of the banner will be 256 square inches. All checkerboards have eight rows and eight columns of squares. How long is the rod you'll need to hang the banner? How many square inches of red cloth will you need to make the banner? Write a sentence explaining how you arrived at your answer.

   S^2 = 256

   S = 16

   The rod should be 16 inches long. You will need 128 square inches of red cloth since half of the total squares are red.

Assignment for Area of Parallelogram

Name: ______________________________

1. Find the area of each parallelogram. Figures are not drawn to scale.

a.

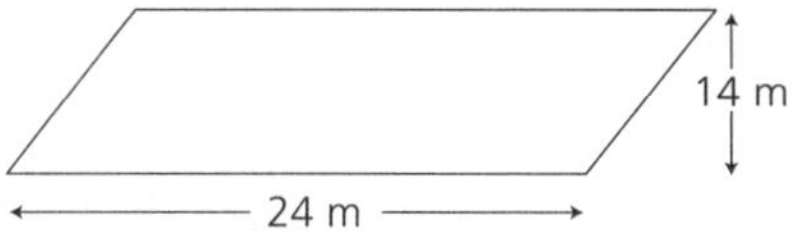

A = 14(24) = 336 m²

b.

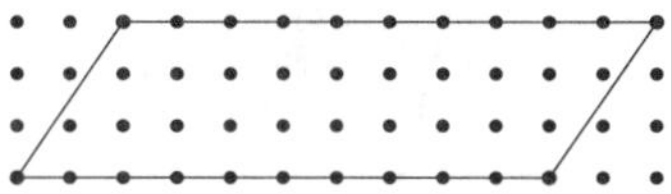

A = 3(10) = 30 square units

c.

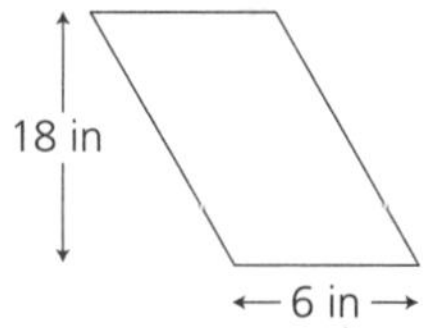

A = 18(6) = 108 in²

d.

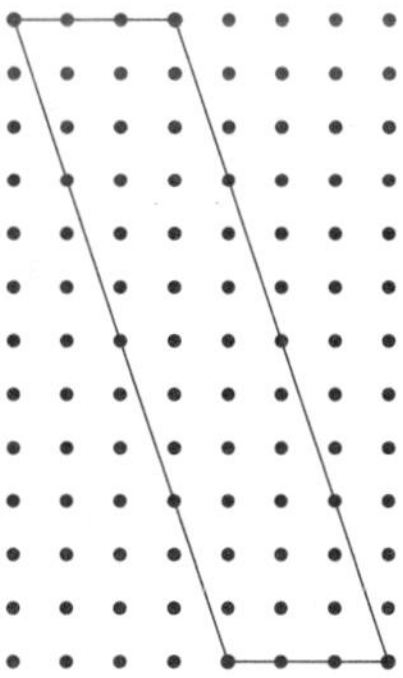

A = 3(12) = 36 square units

e.

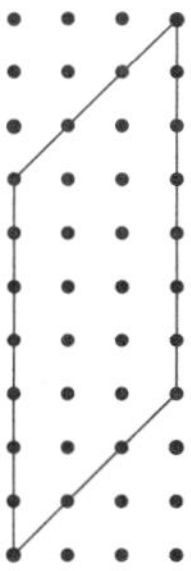

A = 7(3) = 21 square units

2. If the height of a parallelogram is 12 inches and the length of each side is 14 inches, find the area of the parallelogram. Show your work.

   A = 12(14) = 168 in²

3. If the base and height of a parallelogram are equal, must the parallelogram be:

   a. a square? ***No***

   b. a rectangle? ***No***

   Explain your answer.

   In the figure below, both the base and height are equal (3 units), but the figure is neither a square nor a rectangle.

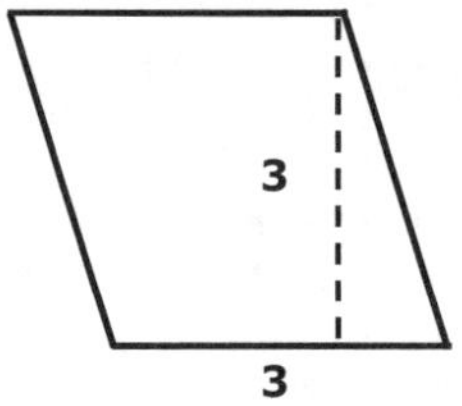

4. If the measures of the base and height of a parallelogram are doubled, will the area also double? Explain your answer.

   $b_2 = 2b$

   $h_2 = 2h$

   $A = bh$

   $A_2 = (2b)(2h) = 4bh$

   No, the area will quadruple.

5. The area of a parallelogram is 234 square feet and the base is 13 feet. Find the height of the parallelogram. Show your work.

   $13h = 234$

   h = 18 feet

   The height of the parallelogram is 18 feet.

6. If the area of a parallelogram is 336 square yards and the height is 32 yards, find the length of the base of the parallelogram. Show your work.

   32b = 336

   b = 10.5 yards

   The base of the parallelogram is 10.5 yards long.

7. The area of a parallelogram is 100 square inches and the measure of the base is twice the height. Find the length of the base and the height of this parallelogram. Show your work.

   b = 2h so

   h(2h) = 100.

   By simplification, $2h^2 = 100$.

   Dividing both sides by 2, $h^2 = 50$.

   Taking the square root of both sides, h ≈ 7.07 inches

   b ≈ 2(7.07) = 14.4 inches

   The height of the parallelogram is 7.07 inches; the base is 14.14 inches long.

8. The area of a parallelogram is 96 square inches and the measure of the base is three times the height. Find the length of the base and the height of this parallelogram. Show your work.

   b = 3h

   h(3h) = 96

   $3h^2 = 96$

   $h^2 = 32$

   h = 5.66 inches

   b = 3(5.66) = 16.98 inches

   The height of the parallelogram is 5.66 inches; the base is 16.98 inches long.

Assignment for Area of Triangles

Name: ______________________________

1. Find the area of both triangles.

   a. 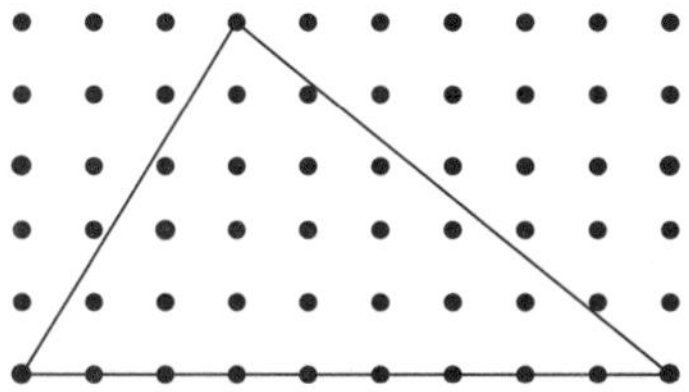

   A = ½ (9)(5) = 22.5

   b. 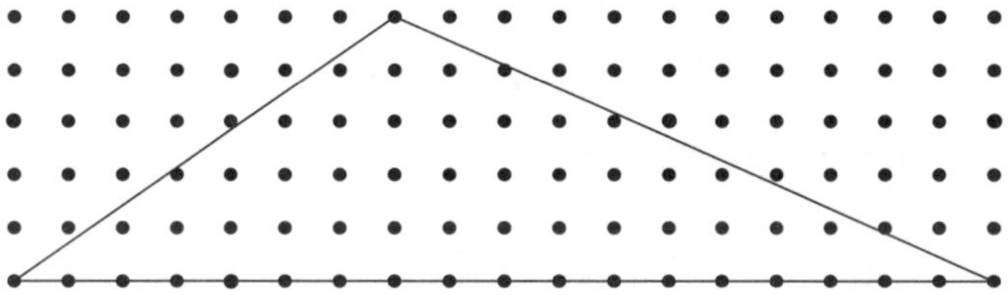

   $A = \frac{1}{2}(18)(5)$

   = 45 square units

   c. If the length of the base of a triangle doubles and the height remains the same, does the area of the triangle double? Explain.

   $A = \frac{1}{2}bh$

   $b_2 = 2b$

   $A_2 = \frac{1}{2}(2b)h$

   $A_2 = bh$

   $A_2 = 2A$

   Yes, the area doubles.

   In (a), the base was 9 and height 5, but in (b) the base was 18 and height 5. The area doubled from (a) to (b). It was 22.5 and became 45.

2. Find the area of each figure.

a. 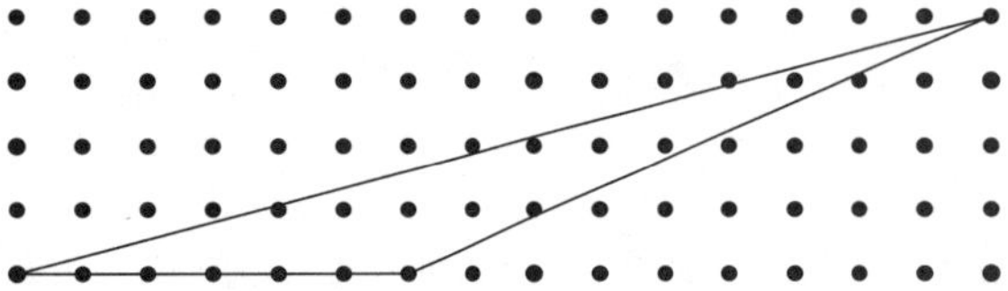

$A = \frac{1}{2}(6)(4)$

$= 12$ square units

b. 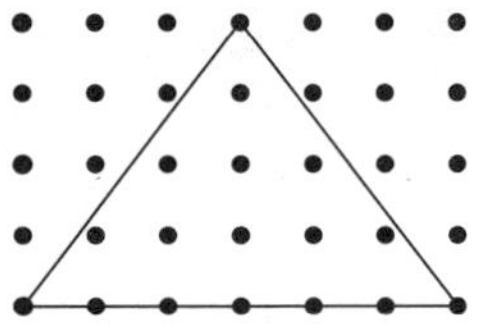

$A = \frac{1}{2}(6)(4)$

$= 12$ square units

c. 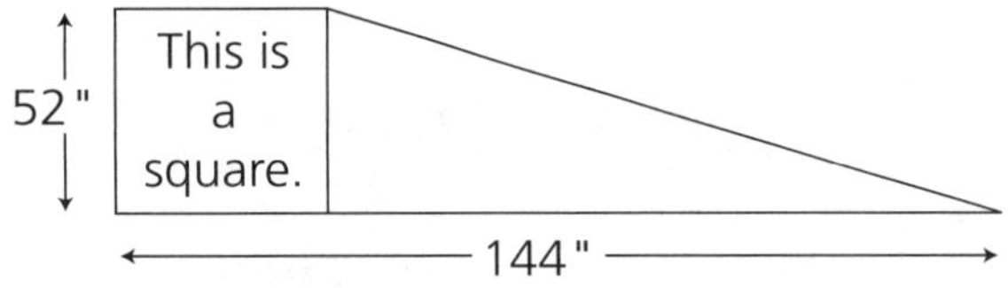

Area $_{square}$***: $A = 52 \bullet 52 = 2704\ in^2$***

Area $_{triangle}$***: $A = \frac{1}{2} \bullet 52 \bullet (144 - 52) = 2392\ in^2$***

Total Area of the figure: $2704 + 2392 = 5096\ inches^2$

3. You are making a kite. If the height of the kite is 36 inches and the widest part of the kite is 24 inches, find the amount of nylon fabric you will need.

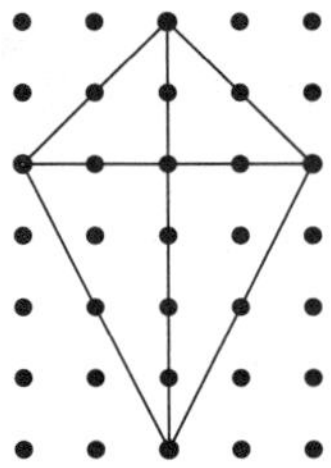

Area of left triangle = $\frac{1}{2}(36)(12)$

$= 216\ in^2$

Total area = $2(216) = 432\ in^2$

4. You are designing a triangular pennant for your school. If you have only 117 square inches of fabric and you want the pennant, when suspended from its base, to hang down 18 inches, what must the length of the base be?

$\frac{1}{2}b(18) = 117$

$9b = 117$

b = 13 inches

The base must be 13 inches.

5. The perimeter of a baseball diamond is 360 feet. (A baseball diamond is actually a square.) How many square feet of grass must be mowed inside the triangle formed by home base, first base, and second base? Write a sentence explaining how you arrived at your answer.

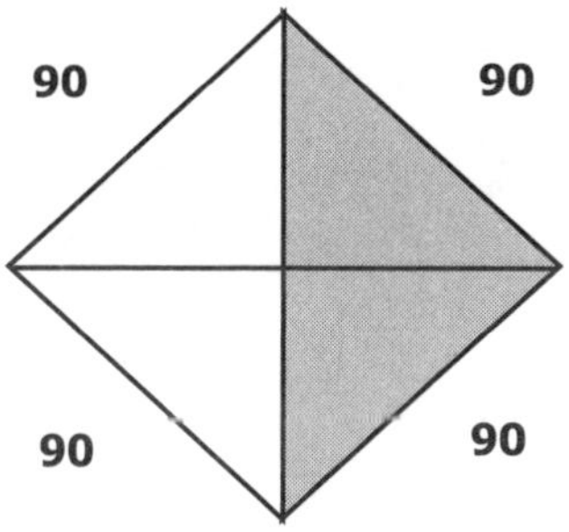

$A = \frac{1}{2}(90)(90)$

$= 4050 \text{ ft}^2$

6. The sail on a boat is triangular and its area is 216 square feet. If the length of the base of the sail is 18 feet, find its height.

The sail is 24 feet high.

$\frac{1}{2}(18)h = 216$

$9h = 216$

$h = 24 \text{ ft}$

7. A triangular window in a beach house is 48 inches in height and contains 1728 square inches of glass. Find the measure of the base of the window.

   $\frac{1}{2}b(48) = 1728$

   24b = 1728

   b = 72 inches

   The base is 72 inches.

Assignment for Area of Trapezoid

Name: ____________________________

1. Find the area of each trapezoid. Figures are not drawn to scale.

a.

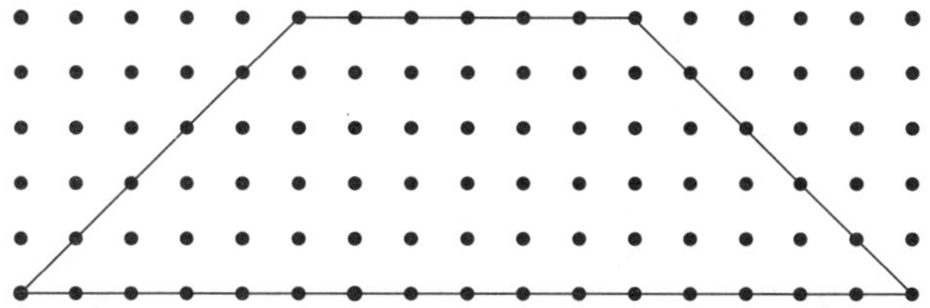

$$A = \frac{1}{2}(6 + 16)(5)$$

$$= \frac{1}{2}(22)(5)$$

$$= 55 \text{ square units}$$

b.

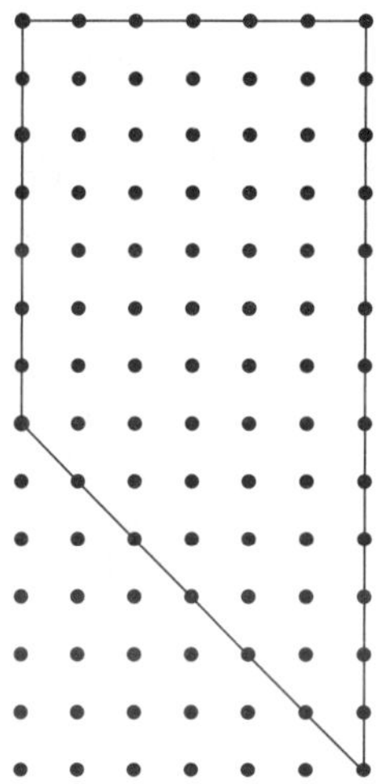

$$A = \frac{1}{2}(7 + 13)(6)$$

$$= \frac{1}{2}(20)(6)$$

$$= 60 \text{ square units}$$

c.

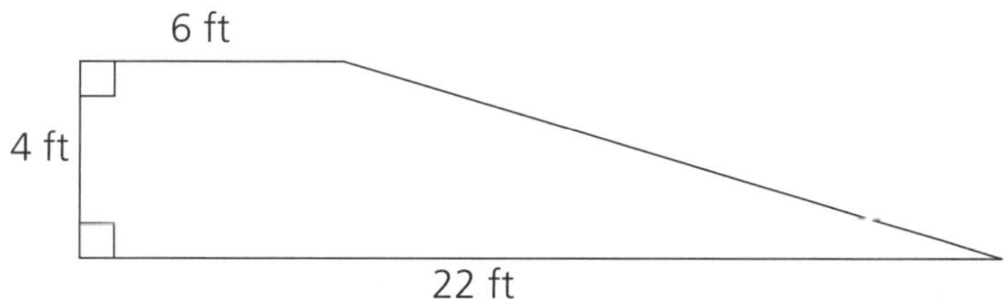

$$A = \frac{1}{2}(6 + 22)(4) = \frac{1}{2}(28)(4) = 56 \text{ ft}^2$$

d.

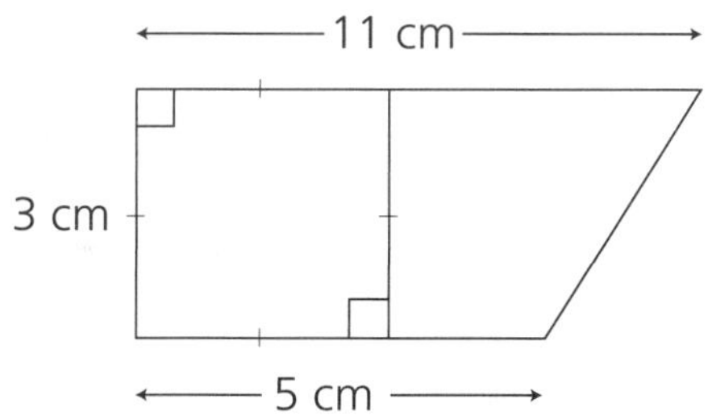

$$A = \frac{1}{2}(5 + 11)(3)$$

$$= \frac{1}{2}(16)(3)$$

$$= 24 \text{ cm}^2$$

2. If the height of a trapezoid is 7 inches, and the bases are 7 inches and 17 inches, find the area of the trapezoid. Show your work.

$$\boldsymbol{A = \frac{1}{2}(7 + 17)(7)}$$

$$\boldsymbol{= \frac{1}{2}(24)(7)}$$

$$\boldsymbol{= 84\ in^2}$$

The area of the trapezoid is 84 square inches.

3. Can the bases of a trapezoid ever be the same length? Explain.

If the bases were the same length, the figure would be a parallelogram, not a trapezoid.

4. If the bases of a trapezoid are 22 feet and 32 feet, and the height is 12 feet, find the area of the trapezoid.

$$\boldsymbol{A = \frac{1}{2}(22 + 32)(12)}$$

$$\boldsymbol{= \frac{1}{2}(54)(12)}$$

$$\boldsymbol{= 324\ ft^2}$$

The area of the trapezoid is 324 square feet.

5. The area of a trapezoid is 209 square yards, and the bases are 15 yards and 23 yards. Find the height of the trapezoid. Show your work.

$$\boldsymbol{\frac{1}{2}(15 + 23)h = 209}$$

$$\boldsymbol{\frac{1}{2}(38)h = 209}$$

$$\boldsymbol{19h = 209}$$

$$\boldsymbol{h = 11\ yards}$$

The height of the trapezoid is 11 yards.

6. The area of a trapezoid is 323 square miles. One base is 26 miles, and the height is 19 miles. Find the length of the second base of the trapezoid. Show your work.

   $\frac{1}{2}(26 + b)(19) = 323$

   $9.5(26 + b) = 323$

   $247 + 9.5b = 323$

   $9.5b = 76$

   b = 8 miles

   The length of the second base is 8 miles.

7. The area of a trapezoid is 150 square meters. The height is 10 meters, and one base is two meters longer than the other base. Find the length of both bases. Show your work.

   One base is 14 meters; the other base is 16 meters.

   $\frac{1}{2}(b + b + 2)(10) = 150$

   $5(2b + 2) = 150$

   $10b + 10 = 150$

   $10b = 140$

   $b = 14$

8. The area of a trapezoid is 252 square feet and the height is 24 feet. Find the length of both bases if the first base is twice the length of the second base. Show your work.

   $\frac{1}{2}(b + 2b)(24) = 252$

   $12(3b) = 252$

   $36b = 252$

   b = 7 ft

   The bases are 7 feet and 14 feet.

Assignment for Polygons

Name: ______________________________

An *apothem* of a regular polygon is a perpendicular segment from the center to any side of the polygon.

If you are finding the area of a regular polygon by dividing it into triangles, the length of an apothem will serve as the height of those triangles.

1. Find the area of each regular polygon. Figures are not drawn to scale.

a. 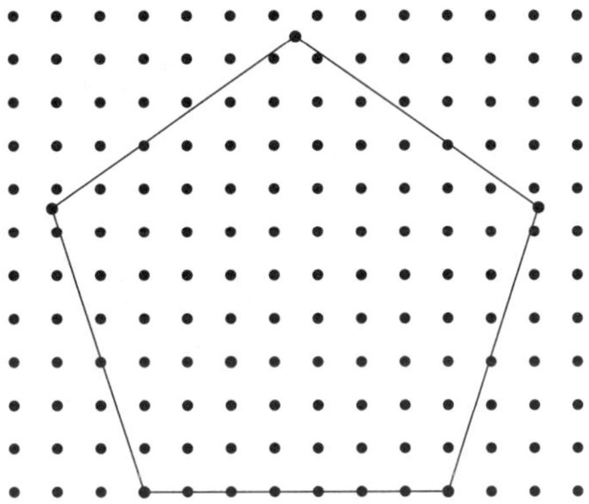

A = ½ (7)(5)(5)

A = 87.5 square units

b. 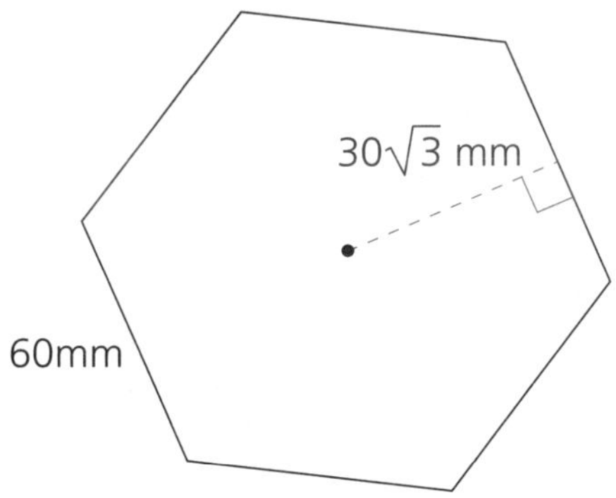

A = ½ (60)($30\sqrt{3}$)(6)

A ≈ 9353.07 mm²

c. 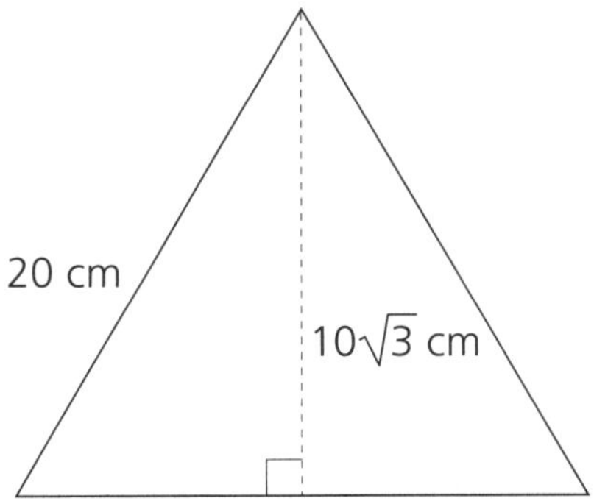

A = ½ (20)($10\sqrt{3}$)

A ≈ 173.21 cm²

2. If the apothem of a regular heptagon is 24.9 inches and each side is 24 inches, find the area of the heptagon. Show your work.

   $A = ½ (24)(24.9)(7)$

   $A = 2091.6\ in^2$

3. If the apothem of a stop sign is 24.1 inches and the length of each side is 20 inches, find the area of the stop sign. Show your work.

   $A = ½ (20)(24.1)(8)$

   $A = 1928\ in^2$

4. If the apothem of a regular decagon is 21.7 meters and the length of each side is 14.1 meters, find the area of the decagon. Show your work.

   $A = ½ (14.1)(21.7)(10)$

   $A = 1529.85\ m^2$

5. If the area of a regular nonagon is 378 square yards and the apothem is 10.5 yards, find the length of each side. Show your work.

   $378 = ½\ s\ (10.5)(9)$

   $378 = 47.25\ s$

   $s = 8$ yards

6. A quilt is made by sewing together interlocking regular hexagons. Each hexagon has an apothem of 1.7 inches and a perimeter of 12 inches. How many regular hexagons will it take to make a quilt that's 6 feet by 8 feet? Show your work.

Area of each hexagon = ½ (2)1.7)(6) = 10.2 in²

Area of the quilt = (72 in)(96 in) = 6912 in²

Number of hexagons needed = $\frac{6912}{10.2}$ ***= 677.65***

You will need approximately 678 hexagons.

7. If the side of a square is 48 meters long, what is:

   a. the area of the square?

   A = 48(48) = 2304 m²

   b. the length of an apothem?

   Apothem is 24 meters, which is half of the length of the side in the case of a square.

8. If the length of a side doubles, how does that affect the area? (Does it double?) How does it affect the length of an apothem?

   If the length doubles, the apothem length will also double. The area will quadruple. (The area doubles once for doubling one dimension and doubles again for the second dimension.)

Assignment for Area of a Circle and Composition Problems

Name: ______________________________

1. Find the area of each figure. (What appears to be a semicircle is just that.)

a. 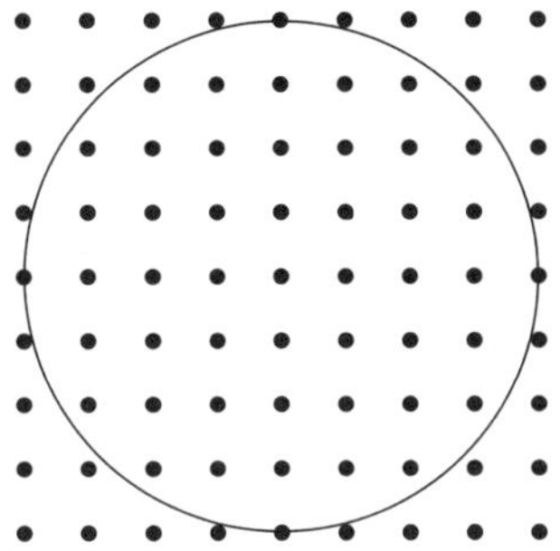

$A = \pi(4)^2$

$A = 16\pi$ square units

or

$A \approx 50.24$ square units

b. 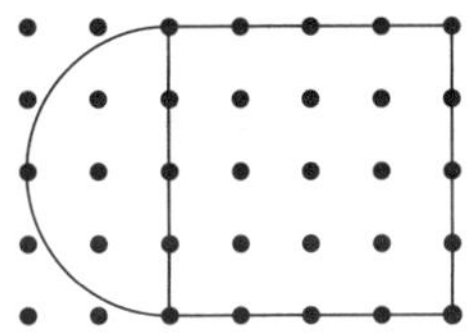

Area of square = 4(4) = 16 sq u

Area of semicircle = $\frac{1}{2}\pi(2)^2 = 2\pi$ sq u

Total Area = $16 + 2\pi \approx 220.28$ sq u

c. 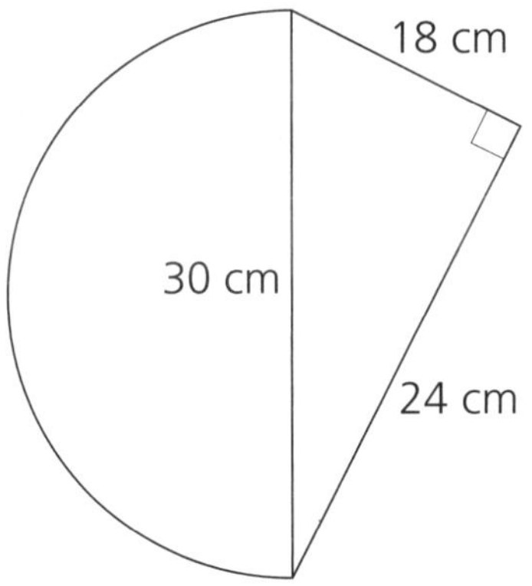

Area of triangle = $\frac{1}{2}(18)(24) = 216\ cm^2$

Area of semicircle = $\frac{1}{2}\pi(15)^2 = 112.5\pi$

Total Area = $216 + 112.5\pi \approx 569.25\ cm^2$

2. A spotlight rotates 360° and emits a light beam that is visible up to 4,500 feet from the light. Find the area covered by the light. Show your work.

$A = \pi(4500)^2$

$A = 20{,}250{,}000\pi\ ft^2$

$A \approx 63{,}585{,}000\ ft^2$

3. A lawn sprinkler spins 360° and sprays water a distance of 3.5 feet. Find the area of lawn the sprinkler will water.

   $\boldsymbol{A = \pi(3.5)^2}$

   $\boldsymbol{A = 12.25\ \pi\ ft^2}$

   $\boldsymbol{A \approx 38.465\ ft^2}$

4. A circular bed of flowers has a circumference of 5 feet. Find the area of the flower bed.

   $\boldsymbol{\pi d = 5}$ $\qquad\qquad$ $\boldsymbol{A = \pi(2.5/\pi)^2}$

   $\boldsymbol{d = 5/\pi}$ $\qquad\qquad$ $\boldsymbol{A = 6.25/\pi}$

   $\boldsymbol{r = 2.5/\pi}$ $\qquad\qquad$ $\boldsymbol{A \approx 1.99\ ft^2}$

5. In the figure below, the area of the square is 25 square feet. The diagonal of the square is $\sqrt{50}$ feet. Find the area of the shaded region.

   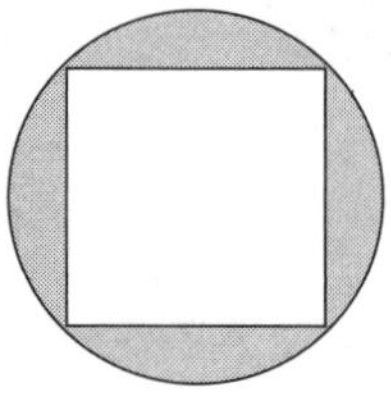

   Area of circle $\boldsymbol{= \pi(\frac{\sqrt{50}}{2})^2 = 12.5\pi ft^2}$

   Area of square = 25 square feet

   Area of shaded region $\boldsymbol{= 12.5\pi - 25 \approx 14.25\ ft^2}$

6. If it takes 7.07 square inches of plastic to manufacture the circular lid for a paper cup, what is the radius of the cup? Show your work.

   $\boldsymbol{\pi r^r}$ **= 7.07 square inches**

   $\boldsymbol{r^2 = 7.07/\pi}$

   $\boldsymbol{r \approx}$ **1.50 inches**

7. In a circular baby pool, 452.39 square feet of water is exposed to the air. To make a cover for the pool you need to know its diameter. Find the diameter of the pool. Show your work.

   πr^2 = 452.39 square feet

   $r^2 = 452.39/\pi$

   $r \approx 12.0$ feet

   diameter = 2 radii = 24.0 feet

8. In this figure, the area of the square is 81 square feet. Find the area of the shaded region. Show your work.

   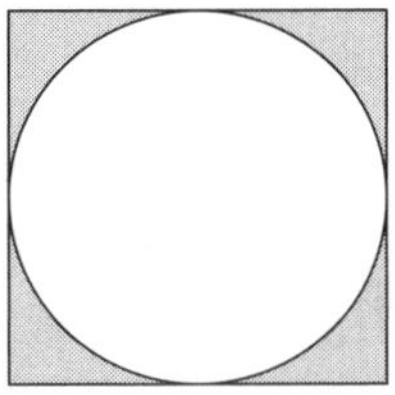

   Area of circle = $\pi(4.5)^2 = 20.25\pi$ square feet

   Area of shaded region = $81 - 20.25\pi \approx 17.415$ feet^2

9. This figure is composed of a semicircle and a square. The area of the square is 196 square feet. Find the perimeter of the figure.

   $s = \sqrt{196}$

   s = 14 feet

   Circumference of semicircle = $\frac{1}{2}\pi(14) = 7\pi \approx 21.98$ feet

   Perimeter $\approx 3(14) + 21.98 = 63.98$ feet

Assignment for Berlin Wall Problem

Name: ______________________________

Review of Squares and Square Roots

Do not use a calculator for this assignment.

1. Memorize the first fifteen perfect squares so you can recognize them when solving problems. It will strengthen your number sense. (Remember that 1^2 means 1 *times* 1, not 1 *plus* 1.)

$1^2 =$ ***1***	$2^2 =$ ***4***	$3^2 =$ ***9***	$4^2 =$ ***16***	$5^2 =$ ***25***
$6^2 =$ ***36***	$7^2 =$ ***49***	$8^2 =$ ***64***	$9^2 =$ ***81***	$10^2 =$***100***
$11^2 =$***121***	$12^2 =$***144***	$13^2 =$***169***	$14^2 =$***196***	$15^2 =$***225***

2. For each square root, write the two consecutive whole numbers between which the square root lies. For example, $\sqrt{32}$ lies between $\sqrt{25}$ and $\sqrt{36}$, so $\sqrt{32}$ lies between 5 and 6.

$\sqrt{60}$: ***7 and 8***	$\sqrt{28}$: ***5 and 6***	$\sqrt{82}$: ***9 and 10***
$\sqrt{174}$: ***13 and 14***	$\sqrt{120}$: ***10 and 11***	$\sqrt{200}$: ***14 and 15***
$\sqrt{150}$: ***12 and 13***	$\sqrt{100}$: ***10***	$\sqrt{46}$: ***6 and 7***

3. Calculate each square root to the nearest tenth. For example, $\sqrt{32}$ is approximately 5.7. You might reason, it is between 5 and 6 ($\sqrt{25}$ and $\sqrt{36}$) and closer to 6 ($\sqrt{36}$), so you know it is more than 5.5. You think it must be around 5.7.

$\sqrt{75}$: ***8.7***	$\sqrt{33}$: ***5.7***	$\sqrt{92}$: ***9.6***
$\sqrt{191}$: ***13.8***	$\sqrt{138}$: ***11.7***	$\sqrt{149}$: ***12.2***
$\sqrt{175}$: ***13.2***	$\sqrt{56}$: ***7.5***	$\sqrt{10}$: ***3.2***

Assignment for Installing a Dish

Name: ______________________________

Review of Squares and Square Roots

Do not use a calculator for this assignment.

1. Memorize these perfect squares, at least up to 25^2, so you can recognize them when solving problems. (Remember that 16^2 means 16 *times* 16, not 16 plus 16.)

   $16^2 =$ ***256*** $17^2 =$ ***289*** $18^2 =$ ***324*** $19^2 =$ ***361*** $20^2 =$ ***400***

   $21^2 =$ ***441*** $22^2 =$ ***484*** $23^2 =$ ***529*** $24^2 =$ ***576*** $25^2 =$ ***625***

   $26^2 =$ ***676*** $27^2 =$ ***729*** $28^2 =$ ***784*** $29^2 =$ ***841*** $30^2 =$ ***900***

2. For each square root, write the two consecutive whole numbers between which the square root lies. For example, $\sqrt{500}$ lies between $\sqrt{484}$ and $\sqrt{529}$, so $\sqrt{500}$ lies between 22 and 23.

   $\sqrt{30}$: ***5 and 6*** $\sqrt{58}$: ***7 and 8*** $\sqrt{77}$: ***8 and 9***

   $\sqrt{204}$: ***14 and 15*** $\sqrt{109}$: ***10 and 11*** $\sqrt{180}$: ***13 and 14***

   $\sqrt{645}$: ***25 and 26*** $\sqrt{886}$: ***29 and 30*** $\sqrt{340}$: ***18 and 19***

3. Calculate each square root to the nearest tenth. For example, $\sqrt{500}$ is approximately 22.4. You might reason, it is between 22 and 23 ($\sqrt{484}$ and $\sqrt{529}$), but it is less than halfway, so you know it is less than 22.5. It must be around 22.4.

   $\sqrt{22}$: ***4.7*** $\sqrt{230}$: ***15.2*** $\sqrt{300}$: ***17.3***

   $\sqrt{191}$: ***13.8*** $\sqrt{1380}$: ***37.1*** $\sqrt{149}$: ***12.2***

   $\sqrt{775}$: ***27.8*** $\sqrt{556}$: ***23.6*** $\sqrt{670}$: ***25.9***

Assignment #1 for Patterns in Right Triangles

Name: ______________________________

1. A plane takes off from an airport and travels 24 miles due west and then 60 miles due north, landing on an island. How far is the airport from the island? Show your work.

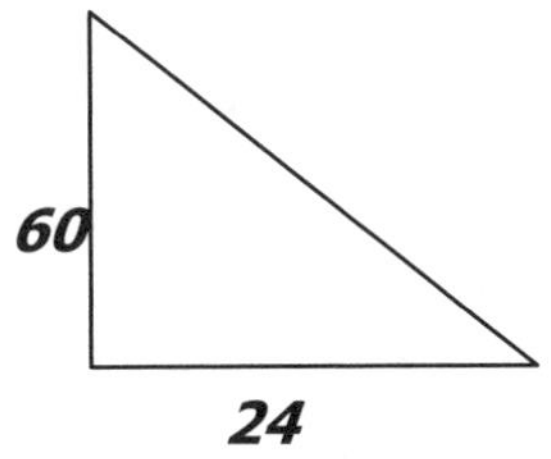

$d^2 = 60^2 + 24^2$

$d^2 = 3600+576$

$d^2 = 4176$

$d \approx 64.62$ ***miles***

2. A helicopter is flying 40 feet above the ocean and spots a man-eating shark directly below. How far is the helicopter from the beach, if the distance from the shark to the beach is 60 feet? Show your work.

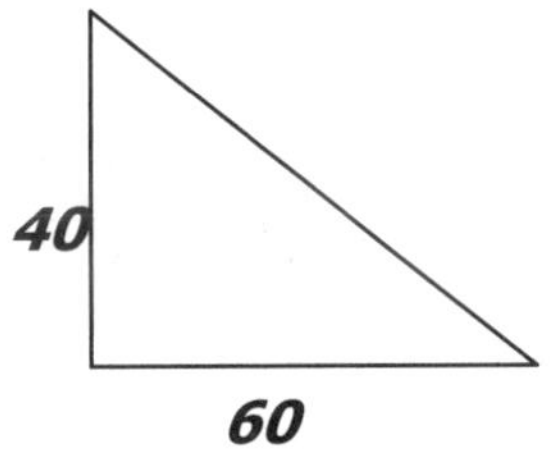

$d^2 = 60^2 + 40^2$

$d^2 = 3600+1600$

$d^2 = 5200$

$d \approx 72.11$ ***feet***

3. The perimeter of a rectangular area rug is 120 feet. Its width is 24 feet. What is the distance between two corners of the rug that do not share a common side? Show your work.

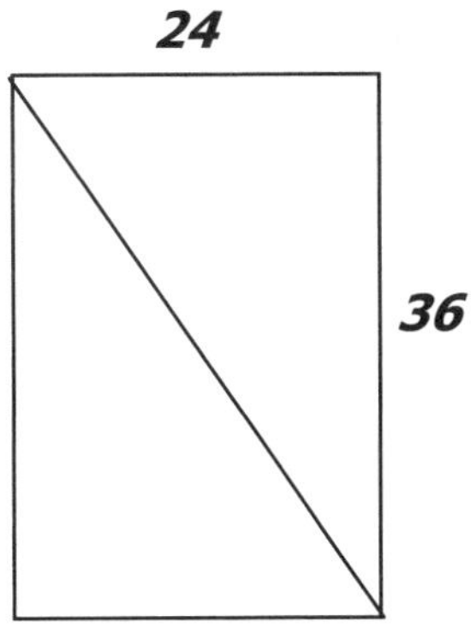

$2(24) + 2l = 120$

$48 + 2l = 120$

$2l = 72$

$l = 36$

$d^2 = 24^2 + 36^2$

$d^2 = 576 + 1296$

$d^2 = 1872$

$d = 43.27$ ***feet***

4. A bird leaves its nest and flies two miles due north, then three miles due east, then four miles due north, and then five miles due east, finally reaching the ocean. How far is its nest from the ocean? Show your work.

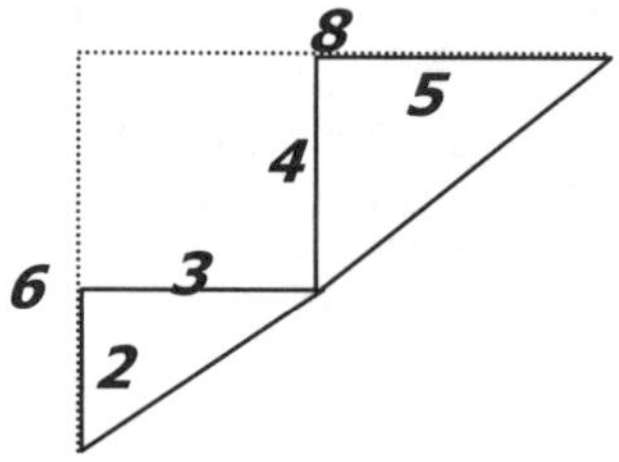

$d^2 = 6^2 + 8^2$

$d^2 = 36+64$

$d^2 = 100$

$d = 10$ ***miles***

5. Two friends leave Pittsburgh at the same time in separate cars. One car travels due east, while the other car travels due north. Both cars travel at 55 miles per hour. Each car has a CB radio with a talking range of 400 miles. In how many hours will they be too far apart to communicate on their radios? Show your work.

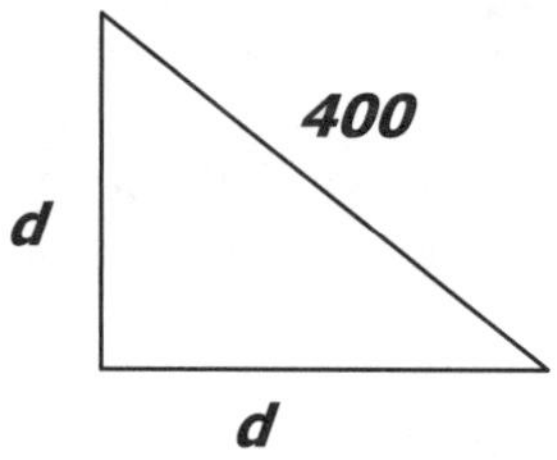

$d^2 + d^2 = 400^2$

$d^2 + d^2 = 160{,}000$

$d^2 = 80{,}000$

$d \approx 292.84$ ***miles***

$55\,t = 282.84$

$t = 5.14$ ***hours***

6. A tree was planted 15.9 feet from a house and eventually grew to a height of 23 feet. During a particularly bad electrical storm, lightening struck the tree, causing the top of the tree to break off 6 feet above the ground. If the severed part of the tree is falling toward the house, will it hit the house? Justify your conclusion.

Since the tree severs 6 feet from the ground, that leaves 17 feet of the tree to fall towards the house. If the house is only 15.9 feet away, the tree will hit the house. There is no need to use the Pythagorean Theorem to deduce this.

7. A plane is 5 miles directly above a house and 42 miles from the Greater Pittsburgh International Airport. How far is the house from the airport? Show your work.

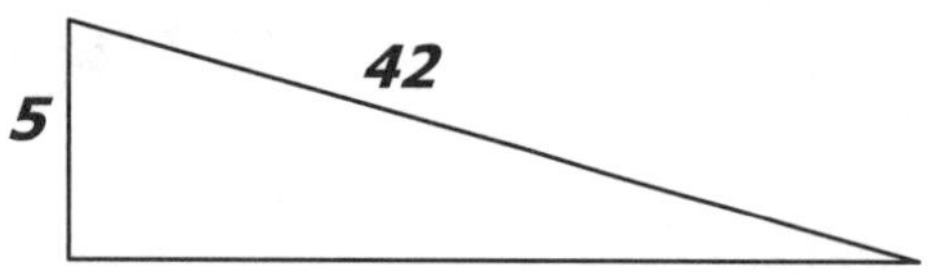

$d^2 + 5^2 = 42^2$

$d^2 + 25 = 1764$

$d^2 = 1739$

$d \approx 41.70$ ***miles***

Assignment #2 for Patterns in Right Triangles

Name: ______________________________

1. This figure is composed of a triangle and a semicircle. Find the area of this figure. Show your work.

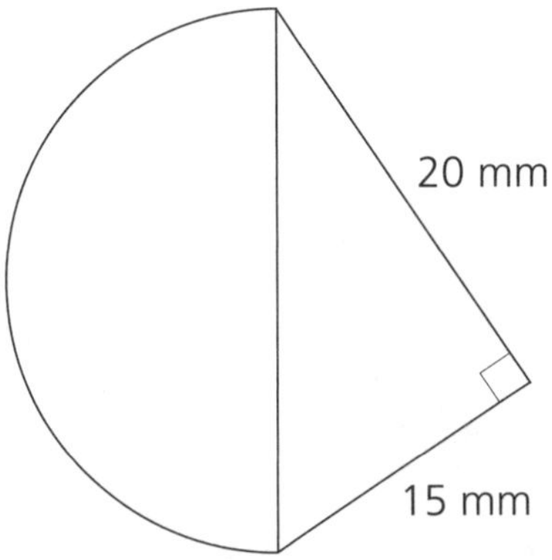

$d^2 = 20^2 + 15^2$

$d^2 = 400 + 225$

$d^2 = 625$

$d = 25$

$r = 12.5$ mm

Area of semicircle $= \frac{1}{2}\pi(12.5)^2 \approx 245.1325$ mm^2

Area of triangle $= \frac{1}{2}(15)(20) = 150$ mm^2

Total Area $= 150 + 245.3125 = 395.3125$ mm^2

2. If this rectangular box is 25 cm long, 20 cm wide, and 10 cm in height, find the distance between point *C* and point *K*. Show your work.

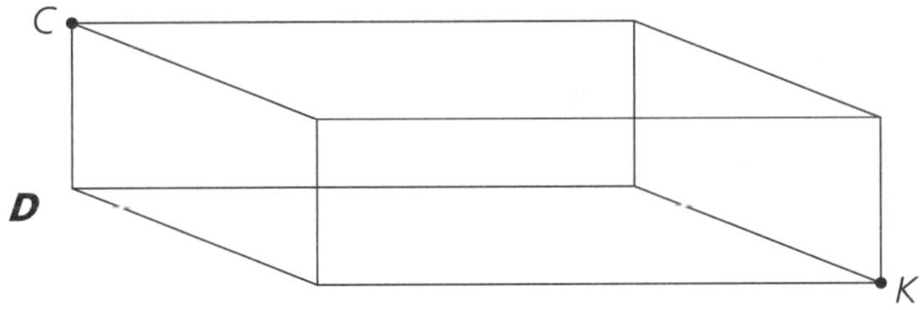

$DK^2 = 20^2 + 25^2$

$DK^2 = 400 + 625$

$DK^2 = 1025$

$DK \approx 32.01$ cm

$CK^2 = 10^2 + 32.01^2$

$CK^2 = 100 + 1024.6401$

$CK^2 = 1124.6401$

$CK \approx 33.54$ cm

3. If a rectangular television screen is 45 cm long and 32 cm wide, what is the diagonal measurement of the screen?

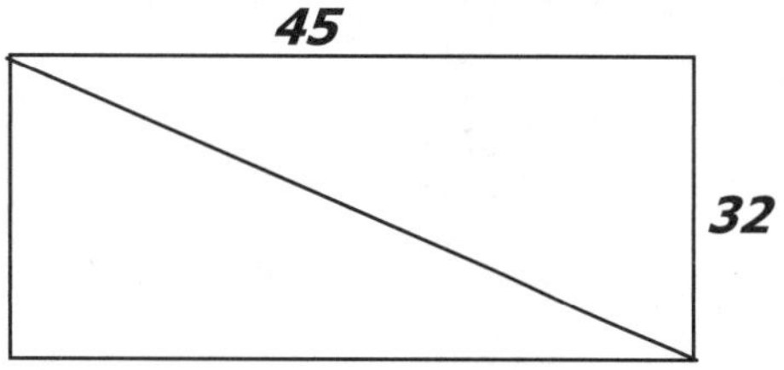

$d^2 = 32^2 + 45^2$

$d^2 = 1024 + 2025$

$d^2 = 3049$

$d \approx 55.22$ cm

4. Will a 15-inch stick fit inside this rectangular box?

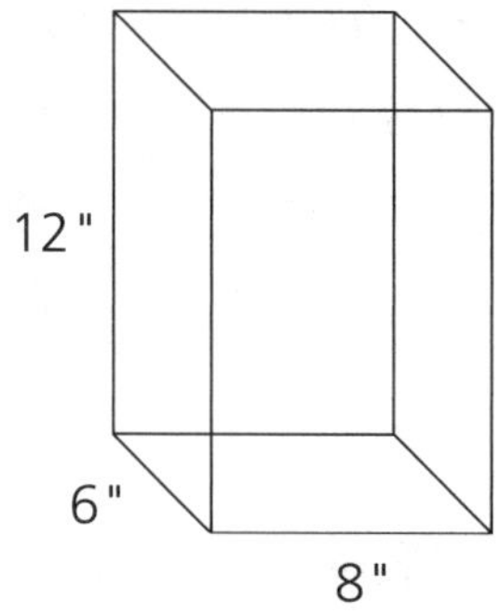

diagonal along bottom face^2 = $6^2 + 8^2$

diagonal along bottom face^2 = 36 + 64

diagonal along bottom face = 10"

Diagonal of box^2 = $10^2 + 12^2$

Diagonal of box^2 = 100 + 144

Diagonal of box ≈ 15.62 inches

Thus, a 15 inch walking stick will fit into the box diagonally with a little room to spare.

5. A boat drops an anchor with a 200-foot chain. The lake is 75 feet deep. How far can the boat drift in any one direction? Show your work.

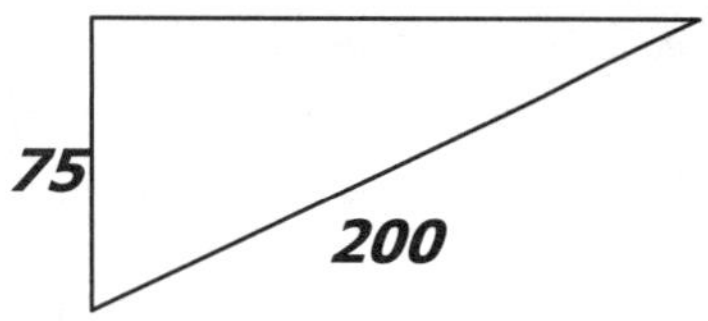

$d^2 + 75 = 200^2$

$d^2 + 5625 = 40000$

$d^2 = 34{,}375$

$d \approx 185.40$ feet

6. Find the length of the longest side of this trapezoid.

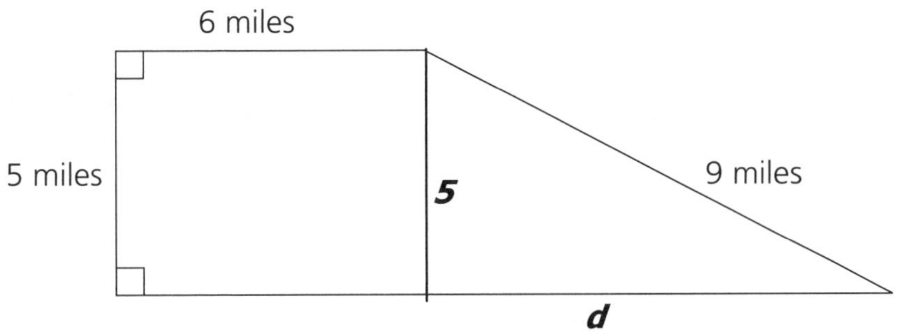

$d^2 + 5^5 = 9^2$

$d^2 + 25 = 81$

$d^2 = 56$

$d \approx 7.48$ miles

Total length = 7.48 +6 = 13.48 miles

Assignment for 45-45-90

Name: ______________________________

Show your work in each problem.

1. If the legs of this isosceles triangle each measure 14 inches, find the length of the hypotenuse.

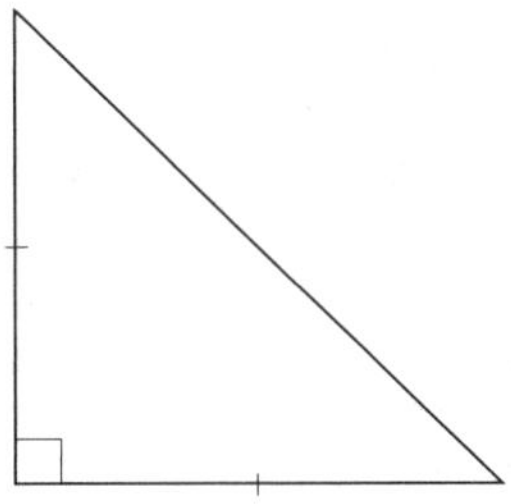

$14\sqrt{2}$ inches is the length of the hypotenuse.

2. Find the value of *c*.

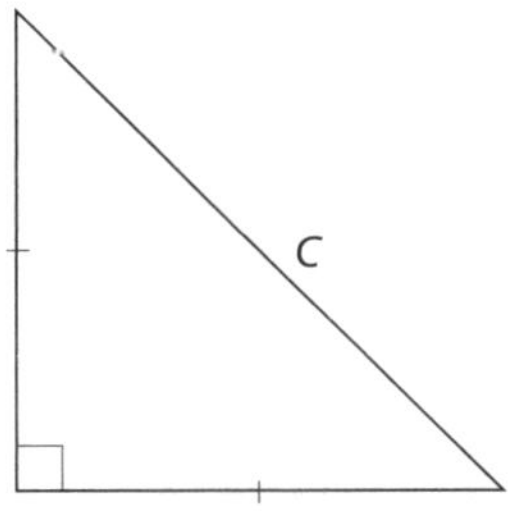

$7\sqrt{2}$

$7\sqrt{2} * \sqrt{2} = 14$

3. The perimeter of this square is 32 cm. Find the length of its diagonal.

side = 32 / 4 = 8 cm

diagonal is $8\sqrt{2}$

4. If the hypotenuse of this isosceles triangle is 20 inches, find the length of the legs.

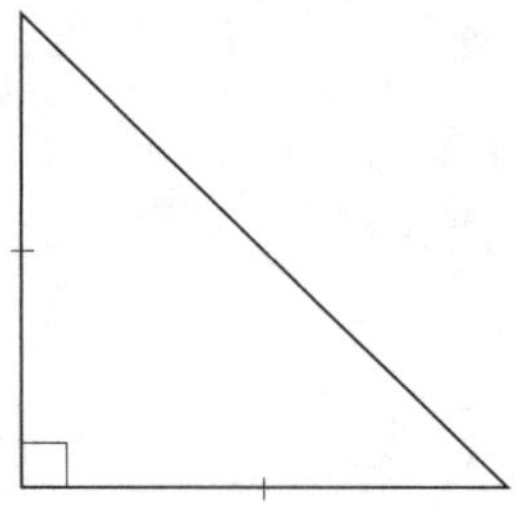

leg = $20/\sqrt{2} \approx 14.14$ inches

5. Find the value of *a*.

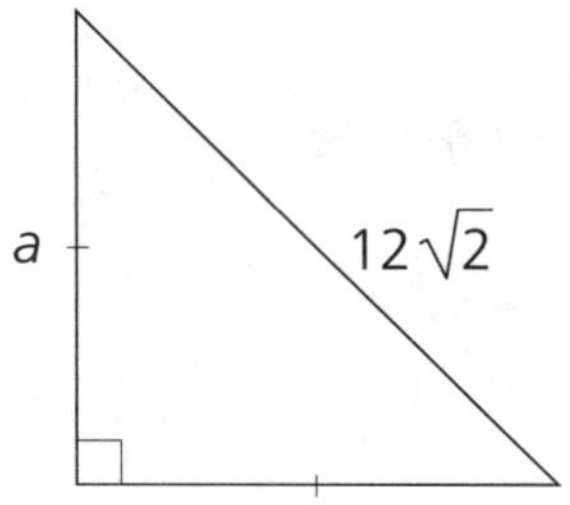

$12\sqrt{2}\,/\sqrt{2} = 12$

6. The length of a diagonal of this square is 36 cm. Find the length of each side.

side = $36/\sqrt{2} \approx 25.46$

7. The diagonal of this square is 12 cm. Find the area.

side = $12/\sqrt{2}$

Area = $(12/\sqrt{2}\,)^2 = 72$ cm^2

8. Find the area of this figure using the information given in the diagram. The figure is composed of a triangle and a semicircle.

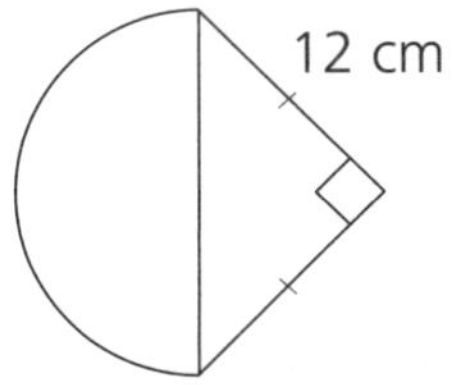

Area of triangle = $\frac{1}{2}$ (12)(12) = 72 cm^2

diameter of circle = $12\sqrt{2}$ cm

radius of circle = $6\sqrt{2}$

Area of semicircle = $\frac{1}{2}\pi(6\sqrt{2})^2 = 36 * \pi cm^2$

Total area = $72 + 36\pi \approx 185.04$ cm^2

9. If the diagonal of the square in the figure below is 60 inches, find the perimeter of the figure. (The figure is composed of a square and a semicircle.)

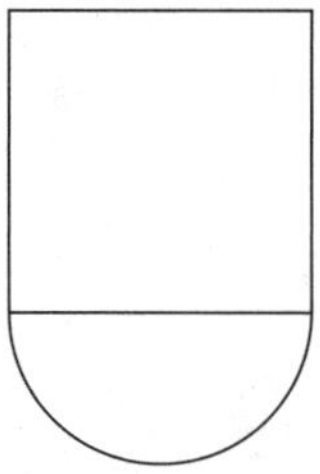

$s = 60/\sqrt{2}$

three sides = $180/\sqrt{2}$

Circumference of semicircle = $\frac{1}{2}\pi(60/\sqrt{2}) = 30/\sqrt{2}$

Perimeter = $180/\sqrt{2} + 30/\sqrt{2} \approx 193.89$ in^2

Assignment for 30-60-90

Name: ______________________________

Show your work in each problem.

1. If the measure of the hypotenuse in this 30-60-90 triangle is 28 meters, find the length of sides *a* and *b*.

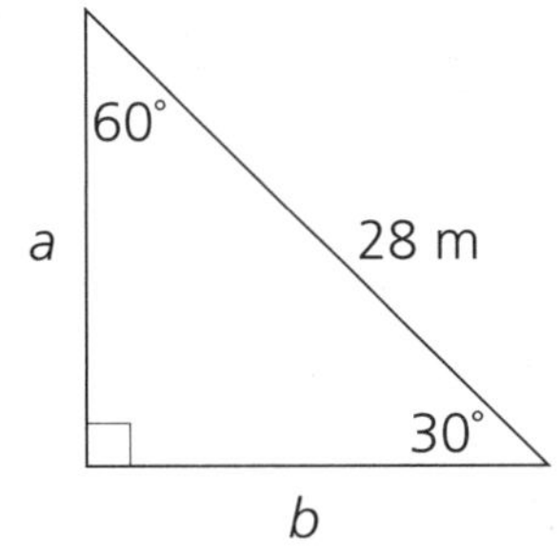

short leg = half the hypotenuse = 14 m

long leg = short leg* $\sqrt{3}$ ***= 14*** $\sqrt{3}$

2. If the measure of the side opposite the 30° angle is 5 feet, find the length of sides *b* and *c*.

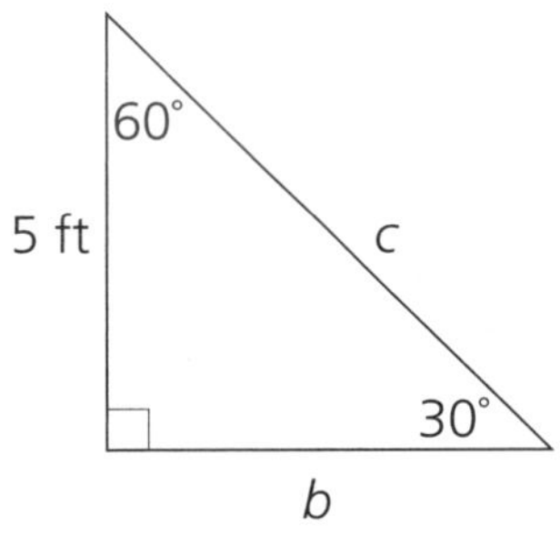

long leg = short leg * $\sqrt{3}$ ***= 5*** $\sqrt{3}$ ***feet***

hypotenuse = short leg *2 = 10 feet

3. If the measure of the side opposite the 60° angle is 8 mm, find the length of sides *a* and *c*.

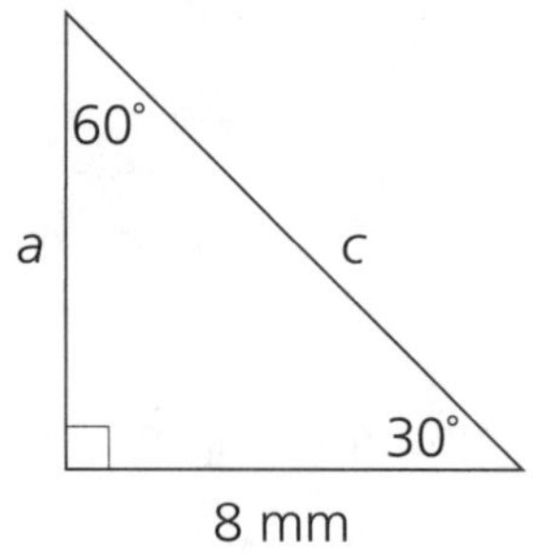

short leg = long leg/$\sqrt{3}$ = 8/$\sqrt{3}$

hypotenuse = short leg * 2 = 2*(8/$\sqrt{3}$) = 16/$\sqrt{3}$

4. A microwave dish is situated on top of a tower. The signal travels from the dish to your house so you can watch TV. The angle of elevation from your house to the tower measures 30° and the distance from your house to the tower is 10 miles. Find the height of the tower and the distance the signal travels.

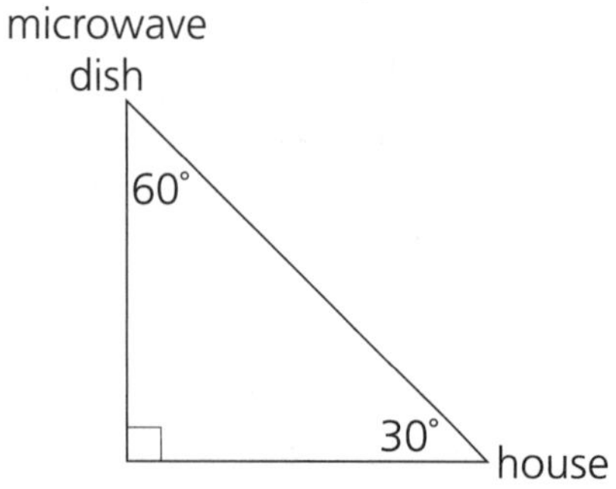

Height of tower is the short leg.

short leg is long leg/$\sqrt{3}$ = 10/$\sqrt{3}$ ≈ 5.77 miles

The height of the tower is 5.77 miles.

hypotenuse is short leg*2 ≈ 11.55 miles

The distance the signal travels is 11.55 miles.

5. The measure of the longest leg in this 30-60-90 triangle is 22 miles. Find the length of the hypotenuse.

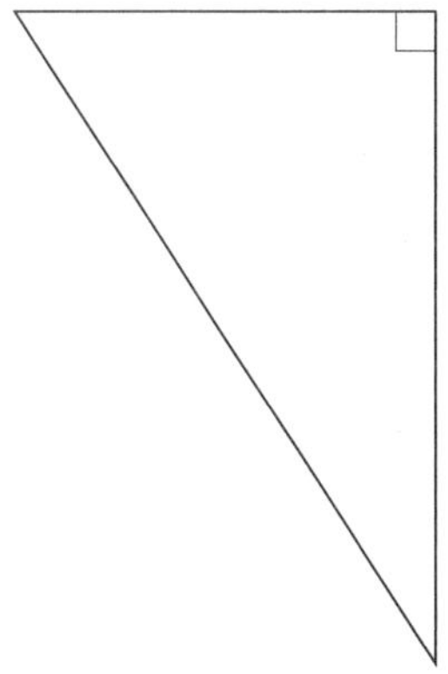

short leg = 22/$\sqrt{3}$ ≈ 12.70

hypotenuse = 2* short leg ≈ 25.4 miles

6. The length of the shortest leg in this 30-60-90 triangle is 13 meters. Find the length of the hypotenuse.

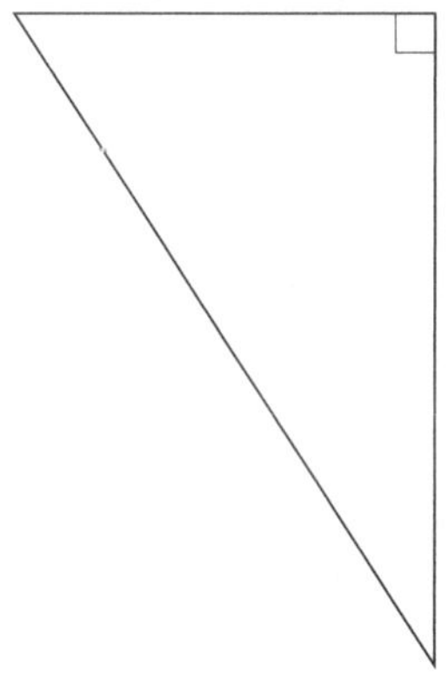

hypotenuse = 2*short leg = 2*13 = 26 meters.

7. Find the perimeter of this trapezoid.

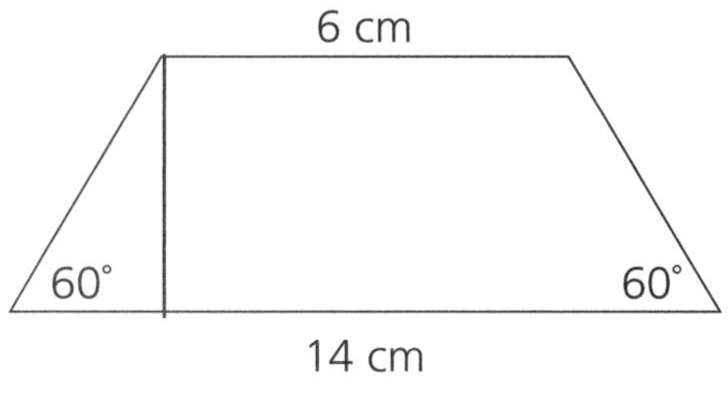

Form a 30-60-90 right triangle as shown.

short leg = (14-6)/2 = 4 cm

hypotenuse = 2*4 = 8 cm

Perimeter = 14 + 6 + 8 + 8 = 36cm

8. Find the area of this triangle.

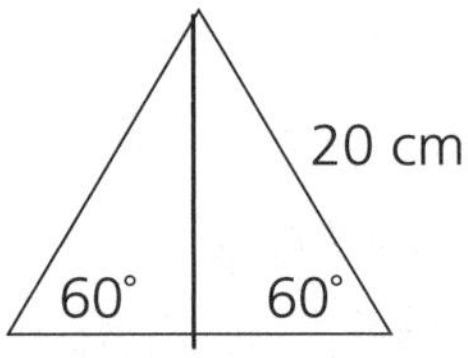

Split the triangle into two 30-60-90 triangles as shown.

base = 20

height = $10\sqrt{3}$

Area = ½ (20)($10\sqrt{3}$) = $100\sqrt{3}$ or about 173.2 cm^2

9. Find the area of this trapezoid.

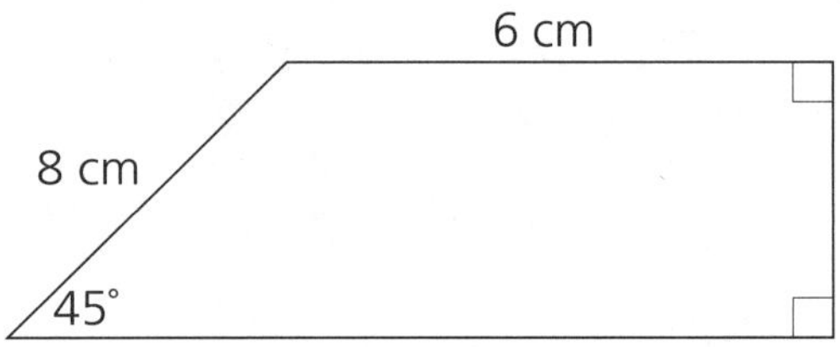

height of the trapezoid = $8/\sqrt{2}$ or about 5.66 cm

length of longer base = 6 + ($8/\sqrt{2}$) or about 11.66 cm

Area of trapezoid ≈ ½ (11.66 + 6)(5.66) = 49.98 cm^2

Assignment for Distance in the Coordinate Plane

Name: ______________________________

1. Use the following grid to graph and connect each given pair of points. Beside each pair of points, write the distance between them. Note that each interval may need to be more than one unit.

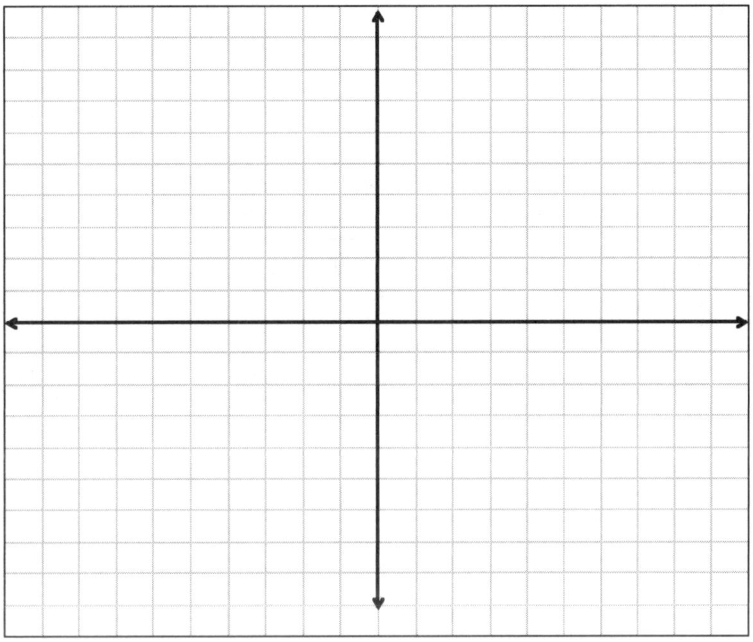

a. (-8, 3) and (-8, 9) Distance: ***6***

b. (-6, 8) and (-1, 8) Distance: ***5***

c. (8, -7) and (-4, -7) Distance: ***12***

d. (8, 8) and (8, -2) Distance: ***10***

2. Describe in a sentence how you could have found the distances in problem 1 from the graph.

Since each pair of points has one coordinate in common, you can count the number of blocks that form the segment connecting the two points.

3. Describe in a sentence how you could have found the distances in problem 1 by using only the numbers given for the coordinates of the points.

   Look at the two coordinates that are different for each set of points, and subtract the smaller from the larger.

4. Use the following grid to graph and connect the given set of three points. Then, find the distances between the points.

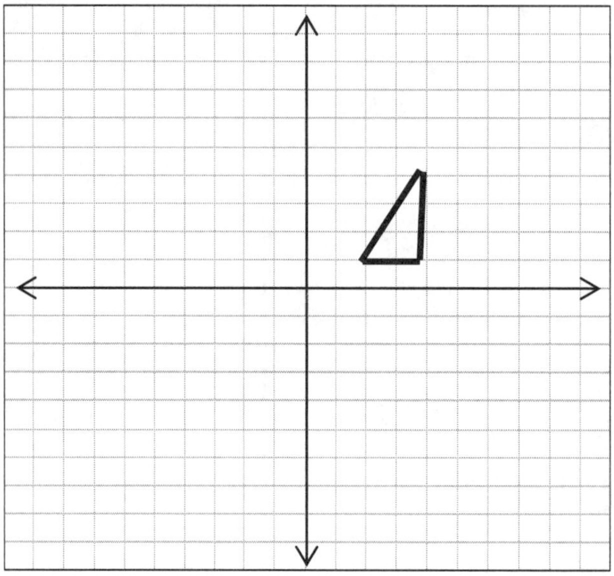

a. (4, 1), (2, 1), and (4, 4) Distance: ***2, 3 and 3.61***

$$\sqrt{2^2 + 3^2} = \sqrt{9 + 4} = \sqrt{13}$$

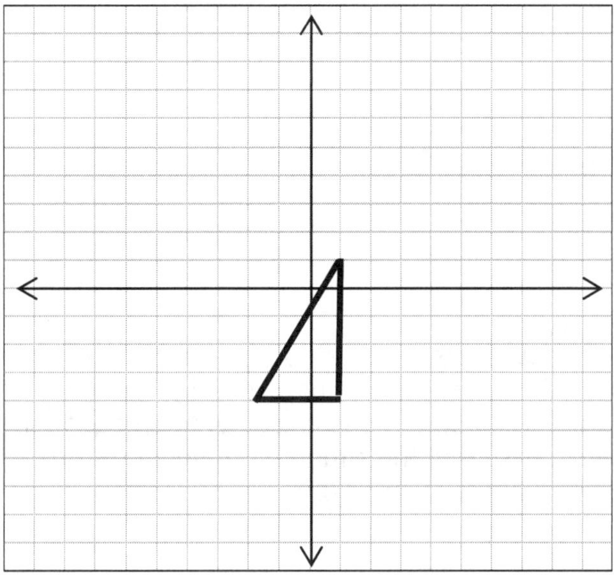

b. (1, -4), (1, 1), and (-2, -4) Distance: ***5, 3, and 5.83***

$$\sqrt{3^2 + 5^2} = \sqrt{9 + 25} = \sqrt{34}$$

5. Describe in a sentence how you found the distances between the points in problem 3.

The three points form a right triangle, so you can use the Pythagorean Theorem to find the required distance.

6. What is the distance between the following pairs of points?

a. (-37, -100) and (14, 0) Distance: $\sqrt{51^2 + 100^2} = \sqrt{2601 + 10000} = \sqrt{12601} = 112.25$

b. (3, 9) and (4, 10) Distance: $\sqrt{1^2 + 1^2} = \sqrt{1 + 1} = \sqrt{2} = 1.414$

c. (-10, -7) and (13, 17) Distance: $\sqrt{23^2 + 24^2} = \sqrt{529 + 576} = \sqrt{1105} = 33.24$

7. Sometimes variables with subscripts are used to represent specific values of a variable. For instance, x_1 can be used to represent a specific value of x. Let (x_1, y_1) and (x_2, y_2) represent two specific points on the Cartesian coordinate plane. Write an equation for the distance between these two points. (Hint: To visualize this situation, draw a right triangle and label the endpoints of the hypotenuse as (x_1, y_1) and (x_2, y_2).)

$$d = \sqrt{(x_2 - x_1)^2 + (y_2 - y_1)^2}$$

Assignment for Designing Roof Trusses

Name: ______________________________

A roof truss was designed with the following specifications.

Height: 5 ft
Span: 24 ft (Half-Span: 12 ft)
Slant Height: 13 ft

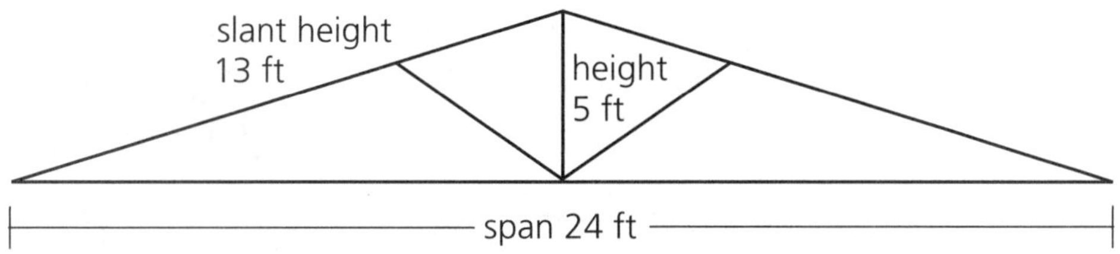

1. Verify that the following side lengths form a right triangle: 5 ft, 12 ft, 13 ft

$5^2 + 12^2 = 13^2$

$25 + 144 = 169$

$169 = 169$

2. A new truss design, proportional to the original, requires that the *half-span* be increased from 12 feet to 14 feet. Define variables for the height and slant height. Then determine values of these variables so that the new truss is proportional to the original dimensions (5 ft, 12 ft, and 13 ft).

$\frac{12}{14} = \frac{5}{h}$

$12h = 70$

$h \approx 5.83$ feet

$\frac{12}{14} = \frac{13}{s}$

$12s = 182$

$s \approx 15.17$ feet

Height: ***5.83 feet*** Slant Height: ***15.17 feet***

3. A new truss design, proportional to the original, requires that the *height* be increased from 5 feet to 6 feet. Define variables for the half-span and slant height. Then determine values of these variables so that the new truss is proportional to the original dimensions (5 ft, 12 ft, and 13 ft).

$\frac{5}{6} = \frac{12}{h}$ $\frac{5}{6} = \frac{13}{s}$

5h = 72 ***5s = 78***

h = 14.4 feet ***s = 15.6 feet***

Half-Span: ***14.4 feet*** Slant Height: ***15.6 feet***

4. A new truss design, proportional to the original, requires that the *height* be *decreased* from 5 feet to 4 feet. Define variables for the half-span and slant height. Then determine values of these variables so that the new truss is proportional to the original dimensions (5 ft, 12 ft, and 13 ft).

$\frac{5}{4} = \frac{12}{h}$ $\frac{5}{4} = \frac{13}{s}$

5h = 48 ***5s = 52***

h = 9.6 feet ***s = 10.4 feet***

Half-Span: ***9.6 feet*** Slant Height: ***10.4 feet***

A kite is flying at the end of 350 feet of string. To determine how high the kite is and how far away the kite is in the horizontal direction, the kite is tied off at ground level.

At a distance of 5 feet along the kite string, the string extends horizontally 4 feet (along the ground), and the end of the string is 3 feet off the ground.

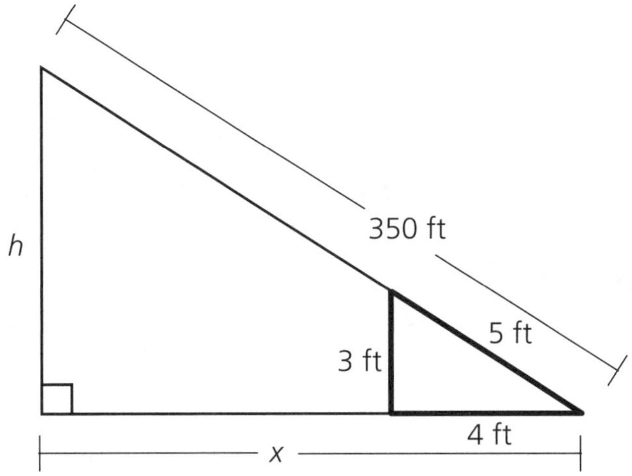

5. Based on the fact that the two triangles in the diagram are similar triangles, use proportions to determine x (the horizontal distance of the kite) and h (the height of the kite).

$\boldsymbol{\frac{5}{350} = \frac{4}{x}}$ $\boldsymbol{\frac{5}{350} = \frac{3}{h}}$

5x = 1400 ***5h = 1050***

x = 280 feet ***h = 210 feet***

Horizontal Distance: ***280 feet*** Height: ***210 feet***

Assignment for ADA Requirements for Wheelchair Ramps

Name: ______________________________

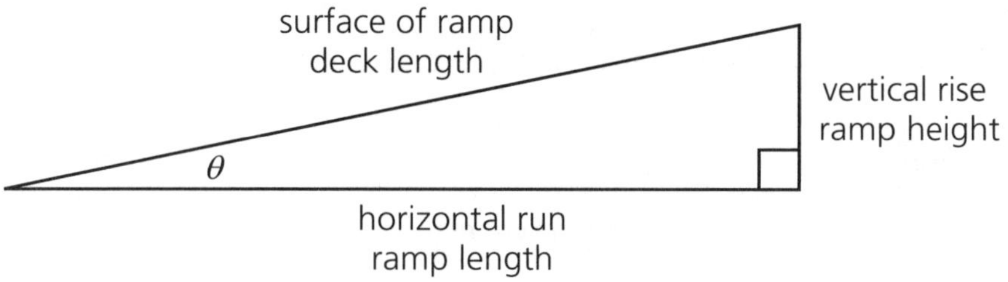

1. Based on the following specifications, determine the slope of each ramp.

   a. Ramp Height: 1.5 feet　　　Ramp Length: 21 feet

   slope = 1.5/21

   slope = 0.07

   Slope: ________________

   b. Ramp Height: 8 inches　　　Ramp Length: 8 feet

   slope = 8/8

   slope =1

   Slope: ________________

Title III, section 4.1.6(3)(a) of the Americans with Disabilities Act states the following.

Curb ramps and interior or exterior ramps to be constructed on sites or in existing buildings or facilities where space limitations prohibit the use of a 1:12 slope or less may have slopes and rises as follows:

(i) A slope between 1:10 and 1:12 is allowed for a maximum rise of 6 inches.

(ii) A slope between 1:8 and 1:10 is allowed for a maximum rise of 3 inches. A slope steeper than 1:8 is not allowed.

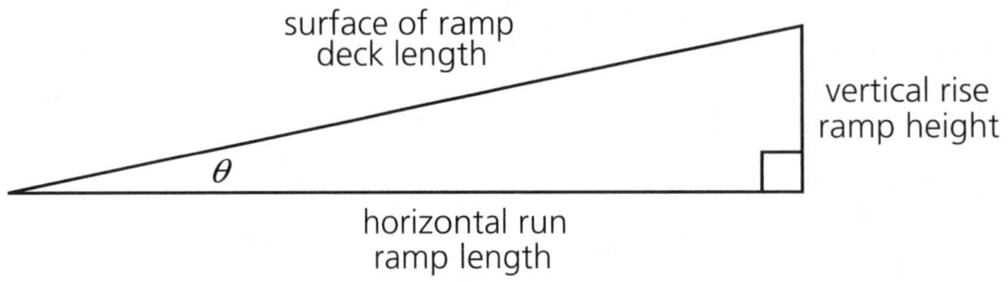

2. If a slope of 1:10 is used to construct a ramp that is 2.5 feet long (horizontal run), define a variable for the ramp height and determine its value in inches.

$$\frac{1}{10} = \frac{h}{2.5}$$

$$h = \frac{2.5}{10}$$

h = 0.25 feet

Ramp Height: ***0.25 feet***

3. If a slope of 1:8 is used to construct a ramp that is 1.5 feet long (horizontal run), define a variable for the ramp height, and determine its value in inches.

$$\frac{1}{8} = \frac{h}{1.5}$$

h = 1.5/8 = 0.1875 feet

Ramp Height: ***0.1875 feet***

4. If a slope of 1:10 is used for a ramp, define a variable for the ramp length and determine its maximum value, based on maximum rise given in part (i) of 4.1.6(3)(a). Report your answer in feet.

$$\frac{1}{10} = \frac{6}{l}$$

l = 60 inches

Maximum Ramp Length: ***60 inches or 5 feet***

5. If a slope of 1:8 is used for a ramp, define a variable for the ramp length, and determine its maximum value, based on the maximum rise given in part (ii) of 4.1.6(3)(a).

   $$\frac{1}{8} = \frac{3}{l}$$

   l = 24 inches

   Maximum Ramp Length: ***24 inches***

Assignment for Tangent Ratio of an Angle

Name: ______________________________

1. A roof truss was designed with the following specifications.

   Half-Span: 10 ft
   Angle of Inclination: 32°

   a. Create a sketch of the truss. Determine the height of the roof truss. Round your answer to the nearest hundredth of a foot.

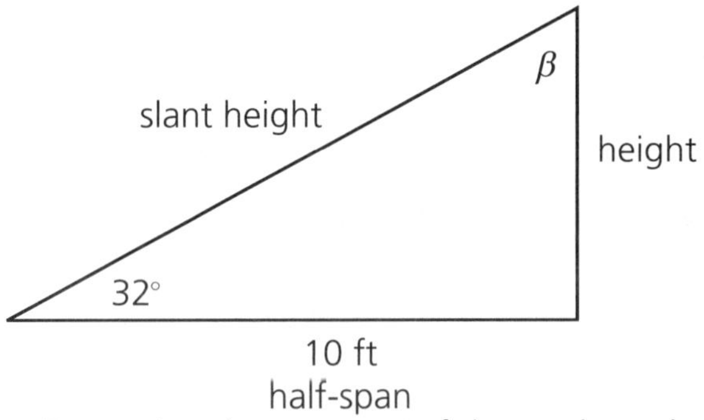

$\tan(32°) = h/10$

$10 * \tan(32°) = h$

$h \approx 6.25$ feet

   b. Determine the measure of the angle at the top of the truss and determine the tangent ratio for that angle.

$\angle\beta = 58°$

$\tan(58°) \approx 1.60$

   c. Determine the slant height.

$s^2 = 10^2 + 6.25^2$

$s^2 = 139.0625$

$s \approx 11.79$ feet

2. Determine θ, the angle of inclination for a wheelchair ramp having each of the following slopes.

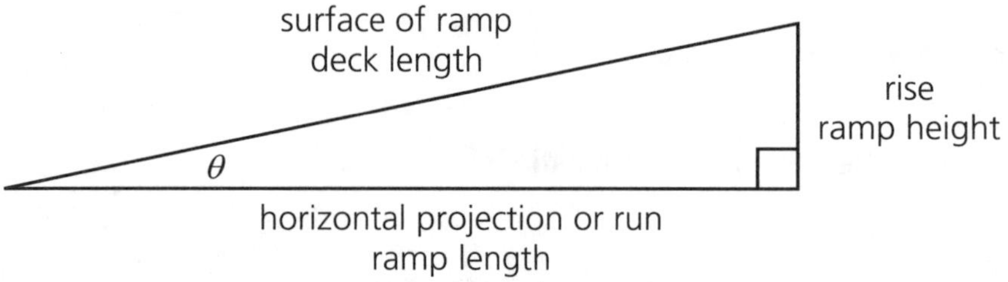

a. Slope = 1:20

$\tan\theta = 1/20$

$\theta = \tan^{-1}(1/20)$

$\theta \approx 2.86°$

Angle of Inclination: ***2.86°***

b. Slope = 1:16

$\tan\theta = 1/16$

$\theta = \tan^{-1}(1/16)$

$\theta \approx 3.58°$

Angle of Inclination: ***3.58°***

c. Slope = 1:10

$\tan\theta = (1/10)$

$\theta = \tan^{-1}(1/10)$

$\theta \approx 5.71°$

Angle of Inclination: ***5.71°***

d. Slope = 1:8

$\tan\theta = (1/8)$

$\theta = \tan^{-1}(1/8)$

$\theta \approx 7.13°$

Angle of Inclination: ***7.13°***

Assignment for Applying an OSHA Regulation

Name: ______________________________

1. OSHA requirements specify that the base of a ladder is to be away from the wall a distance of one-fourth the working length of the ladder. For the ladder given in each of the following problems:

   - Determine the actual distance the base should be placed from the wall
   - Determine how high the ladder reaches up the wall
   - Sketch and label a diagram depicting the situation

   a. A single-section 12-foot ladder (working length: 12 feet)

   b = (¼)(12) = 3 feet

   $3^2 + h^2 = 12^2$

   $9 + h^2 = 144$

   $h^2 = 135$

   $h \approx 11.62$ ***feet***

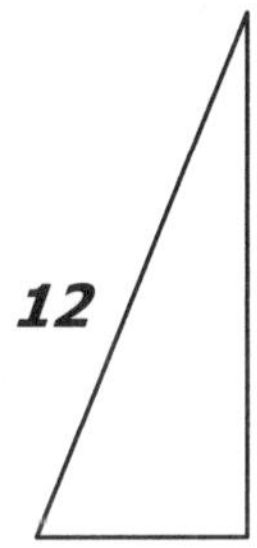

   b. A two-section 40-foot ladder (working length: 36 feet)

   b = (¼)(36) = 9 feet

   $9^2 + h^2 = 36^2$

   $81 + h^2 = 1296$

   $h^2 = 1215$

   $h \approx 34.86$ ***feet***

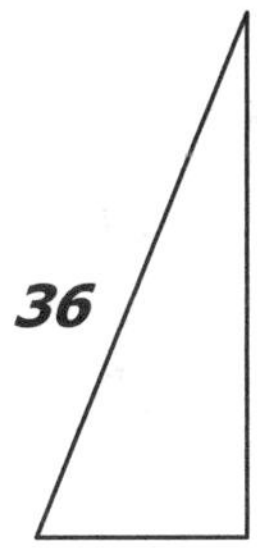

OSHA requirement 1910.29(f)(3) regarding scaffolds, states:

The slope of the steps section shall be a minimum of fifty-five (55) degrees and a maximum of sixty (60) degrees, measured from the horizontal.

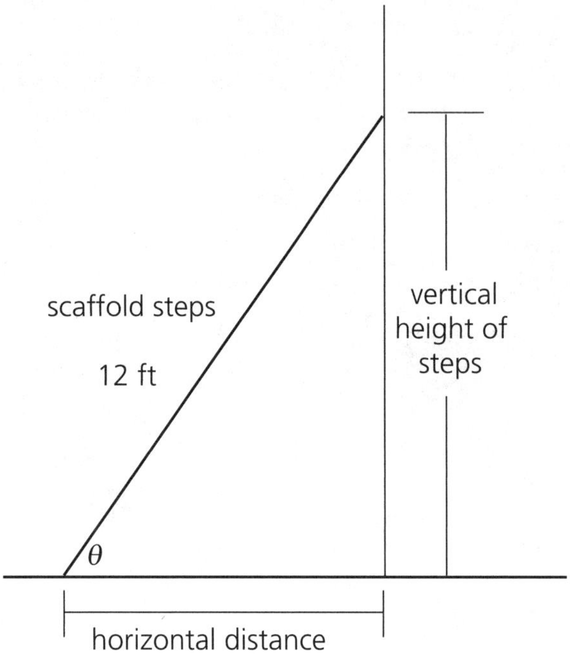

2. If the length of the steps section is 12 feet, determine the vertical height and horizontal distance of the steps if the slope of the steps section is:

   a. the minimum slope (55 degrees)

   sin 55° = y/12 ***cos 55°= x/12***

   12*sin 55° = y ***12*cos 55° = x***

   y ≈ 9.83 feet ***x ≈ 6.88 feet***

Vertical Height: ***9.83 feet*** Horizontal Distance: ***6.88 feet***

b. the maximum slope (60 degrees)

sin 60° = y/12

12*sin 60° = h

y ≈ 10.39 feet

cos 60° = x/12

12*cos 60° = x

x = 6 feet

Vertical Height: ***10.39 feet*** Horizontal Distance: ***6 feet***

3. Determine all the unknown parts of the following roof truss.

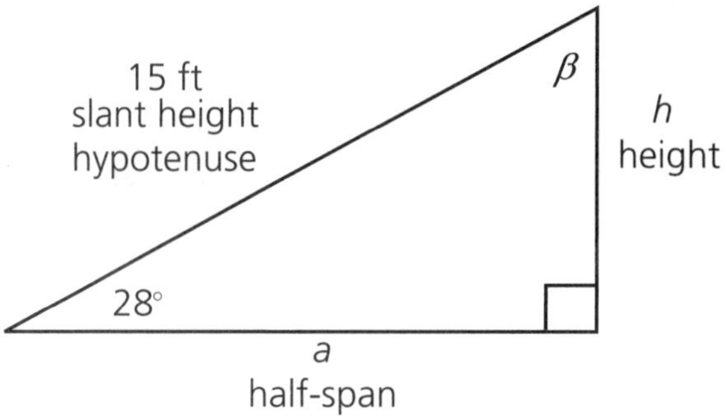

sin 28°= h/15

15*sin28° = h

h ≈ 7.04 feet

cos 28° = a/15

15*cos 28° = a

a ≈ 13.24 feet

Half-Span	Height	Slant Height	Angle α	Angle β
13.24 ft	***7.04 ft***	15 ft	28°	***62°***

Assignment for Sine and Cosine of an Angle

Name: ______________________________

1. In the following square, the diagonal creates an isosceles right triangle.

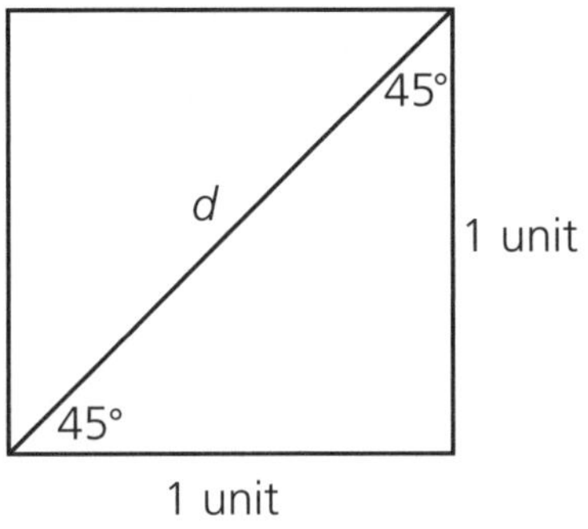

a. Determine the *exact* length of the diagonal. (This means to leave your answer in radical form, rather than using your calculator to approximate the answer.)

$d = \sqrt{2}$

b. Note that the diagonal is also the hypotenuse of the isosceles triangle. Determine the *exact* values for the sine, cosine, and tangent of a 45° angle. (This means to use the side-length ratios for sine, cosine, and tangent, rather than evaluating with your calculator.)

$\sin 45° = \frac{1}{\sqrt{2}}$ $\cos 45° = \frac{1}{\sqrt{2}}$

$\tan 45° = 1$

2. The perpendicular bisector drawn in the following equilateral triangle creates a 30-60-90 degree right triangle.

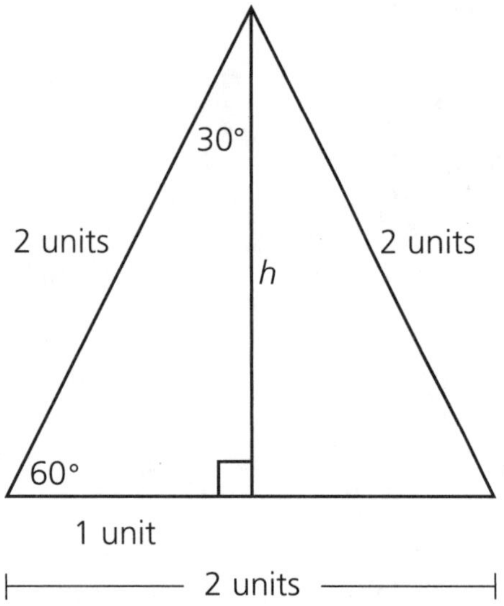

   a. Determine the *exact* length of the perpendicular bisector. (This means to leave your answer in radical form, rather than using your calculator to approximate the answer.)

   $\boldsymbol{h = \sqrt{3}}$

   b. Note that the perpendicular bisector is a leg of the 30-60-90 degree right triangle. Determine the *exact* values for the sine, cosine, and tangent of both a 30° angle and a 60° angle.

   $\sin 30° = \boldsymbol{\frac{1}{2}}$ $\qquad$ $\sin 60° = \boldsymbol{\frac{\sqrt{3}}{2}}$

   $\cos 30° = \boldsymbol{\frac{\sqrt{3}}{2}}$ $\qquad$ $\cos 60° = \boldsymbol{\frac{1}{2}}$

   $\tan 30° = \boldsymbol{\frac{1}{\sqrt{3}}}$ $\qquad$ $\tan 60° = \boldsymbol{\sqrt{3}}$

3. Answer the following questions based on the exact values of the sine, cosine, and tangent for 30°, 45°, and 60° angles.

If the hypotenuse of a 45-45-90 degree right triangle is 20 feet long, what are the leg lengths?

sin45° = a/20

$$\frac{1}{\sqrt{2}} = \frac{a}{20}$$

$$a = \frac{20}{\sqrt{2}}$$

If the legs of a 45-45-90 degree right triangle are 13 cm long, what is the length of its hypotenuse?

sin45° = 13/x

$$\frac{1}{\sqrt{2}} = \frac{13}{x}$$

$$x = 13\sqrt{2}$$

If the hypotenuse of a 30-60-90 degree right triangle is 7 inches long, what are the lengths of the two legs?

sin 30° = a/7

$$\frac{1}{2} = \frac{a}{7}$$

a = 7/2 = 3.5 inches

sin 60°= b/7

$$\frac{\sqrt{3}}{2} = \frac{b}{7}$$

$$b = \frac{7\sqrt{3}}{2} = 3.5\sqrt{3} \approx 6.06 \text{ inches}$$

If the longer leg of a 30-60-90 degree right triangle is 12 feet long, what are the lengths of the hypotenuse and the shorter leg?

$\tan 30° = a/12$

$$\frac{1}{\sqrt{3}} = \frac{a}{12}$$

$a = 12/\sqrt{3} \approx 6.93$ feet

$\sin 60° = 12/c$

$$\frac{\sqrt{3}}{2} = \frac{12}{c}$$

$c = \frac{24}{\sqrt{3}} \approx 13.86$ feet

4. *Solve* the following right triangles. (*Solving* a right triangle means to find all the unknown sides and angles.)

a. 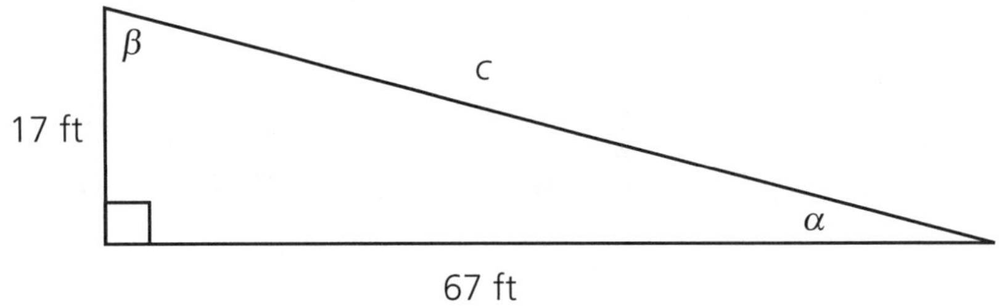

$c^2 = 17^2 + 67^2$

$c^2 = 4778$

$c \approx 69.12$ feet

$\tan\alpha = 17/67$

$\alpha = \tan^{-1}(17/67)$

$\alpha \approx 14.24°$

$\beta = 90° - 14.24°$

$\beta = 75.76°$

c = ***69.12 feet*** α = ***14.24°*** β = ***75.76°***

b.

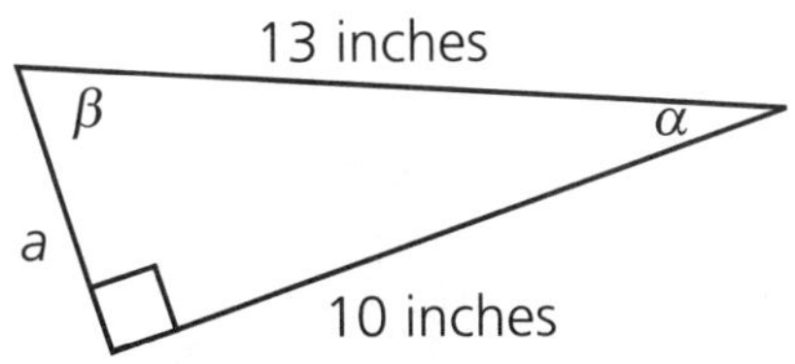

$10^2 + a^2 = 13^2$ $\cos\alpha = 10/13$ $\beta = 90° - 39.72°$

$a^2 = 169 - 100$ $\alpha = \cos^{-1}(10/13)$ $\beta = 50.28°$

$a^2 = 69$ $\alpha = 39.72°$

$a \approx 8.31$ in

a = **8.31 inches** α = **39.72°** β = **50.28°**

c.

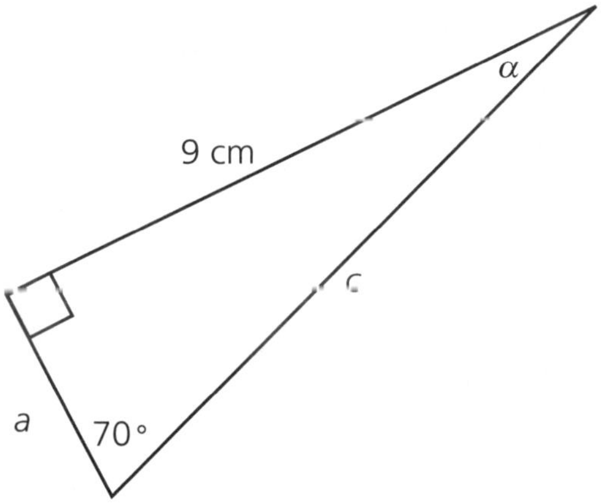

$\alpha = 90° - 70°$ $\sin 70° = 9/c$ $\tan 70° = 9/a$

$\alpha = 20°$ $c*\sin 70° = 9$ $a*\tan 70° = 9$

$c = 9/(\sin 70°)$ $a = 9/\tan 70°$

$c \approx 9.58$ cm $a \approx 3.28$ cm

a = **3.28 cm** c = **9.58 cm** α = **20°**

d.

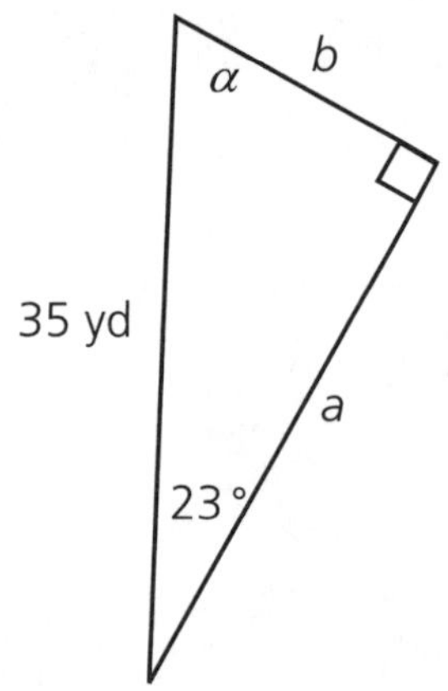

α = 90° - 23°	***sin 23° = b/35***	***cos 23°= a/35***
α = 67°	***35*sin 23° = b***	***35*cos 23° = a***
	b ≈ 13.68 yards	***a ≈ 32.22 yards***

a = ***67°*** b = ***13.68 yards*** α = ***32.22 yards***

Assignment for Angles of Elevation and Depression

Name: ______________________________

1. With its radar, an aircraft spots another aircraft 8,000 feet away at a 12° angle of depression.

   a. Sketch a figure for this problem situation. In the right triangle that is formed, label the angle of depression, the side adjacent to this angle, the side opposite the angle of depression, and the hypotenuse.

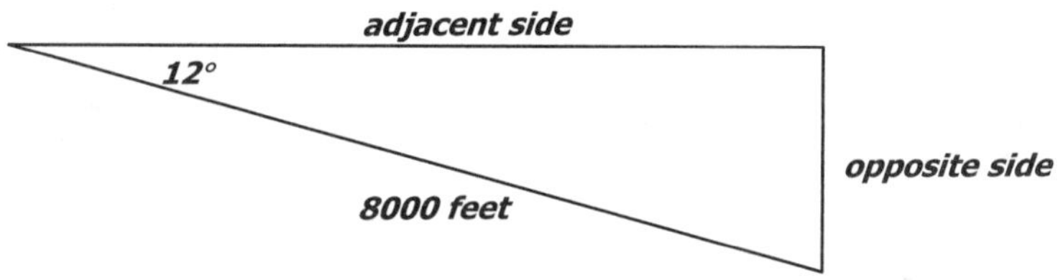

   b. Determine the vertical and horizontal separation of the two aircraft.

$\sin 12° = y/8000$	$\cos 12° = x/8000$
$y = 8000 * \sin 12°$	$x = 8000 * \cos 12°$
$y \approx 1663.29$ feet	$x \approx 7825.18$ feet

2. When a space shuttle returns from a mission, the angle of its descent to the ground from the final 10,000 feet above the ground is between 17° and 19° with the horizontal.

   a. Sketch a figure for this problem situation. In the right triangle that is formed, label the angle of depression, the side adjacent to this angle, the side opposite the angle of depression, and the hypotenuse.

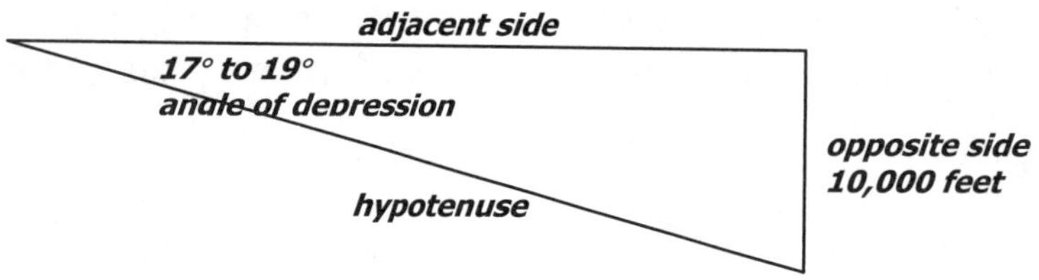

   b. Based on the fact that the angle of depression is between 17° and 19°, determine the minimum and maximum horizontal distance between the landing site and where the descent begins.

MAXIMUM

tan 17° = 10,000/x

x * tan 17° = 10,000

x = 10,000/tan 17°

x = 32,708.53 feet

MINIMUM

tan 19°= 10,000/x

x * tan 19° = 10,000

x = 10,000/tan 19°

x = 29,042.11 feet

Assignment #1 for Exploring Quadratic Equations

Name: ______________________________

Enter each equation into your graphing calculator and graph it.

- Fill in the table and sketch the graph
- Find the highest (or lowest) point on the graph
- Find the x-intercepts and the y-intercept

$y = x^2$

x	y
-2	***4***
-1	***1***
0	***0***
1	***1***
2	***4***

highest/lowest point

(***0*** , ***0***)

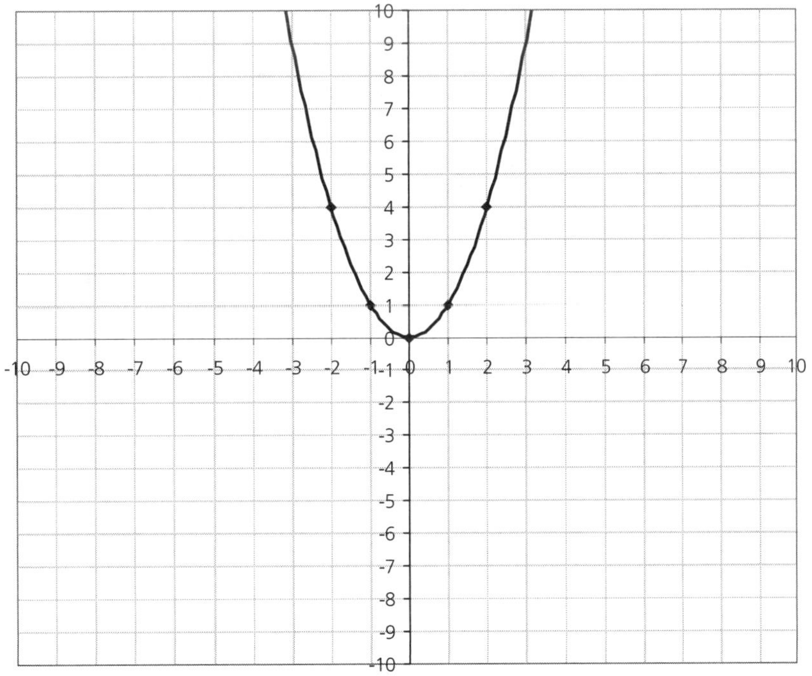

x-intercepts: ***(0,0)***

y-intercept: ***(0,0)***

$y = x^2 + 2$

x	y
-2	***6***
-1	***3***
0	***2***
1	***3***
2	***6***

highest/lowest point

(***0*** , ***2***)

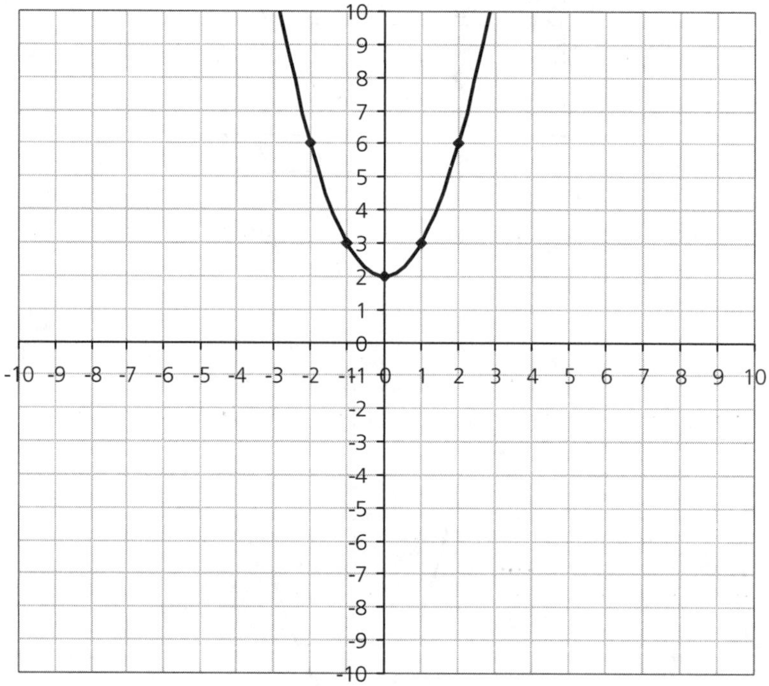

x-intercepts: ***none***

y-intercept: ***(0,2)***

Assignment #2 for Exploring Quadratic Equations

Name: ____________________________

Enter each equation into your graphing calculator and graph it.

- Fill in the table and sketch the graph
- Find the highest (or lowest) point on the graph
- Find the x-intercepts and the y-intercept

$y = x^2 + 3$

x	y
-2	***7***
-1	***4***
0	***3***
1	***4***
-2	***7***

highest/lowest point

(***0*** , ***3***)

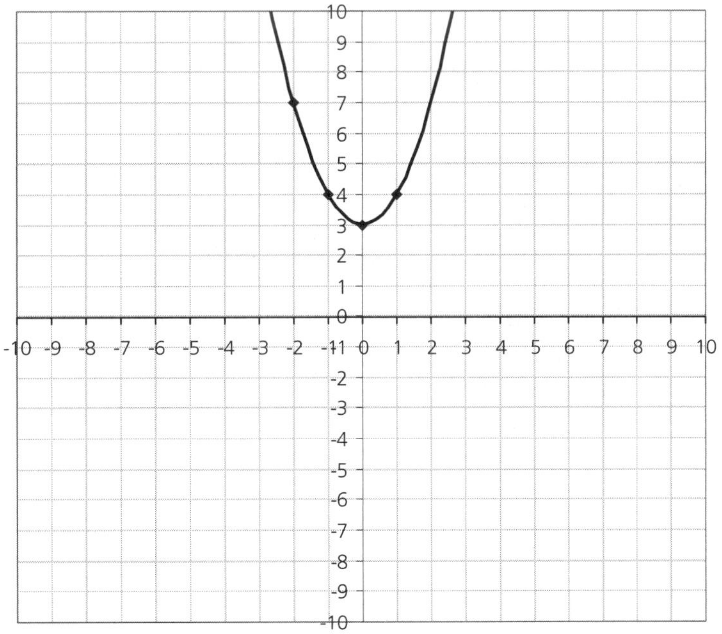

x-intercepts: ***none***

y-intercept: ***(0,3)***

$y = x^2 - 5$

x	y
-2	***-2***
-1	***-4***
0	***-5***
1	***-4***
2	***-1***

highest/lowest point

(***0*** , ***-5***)

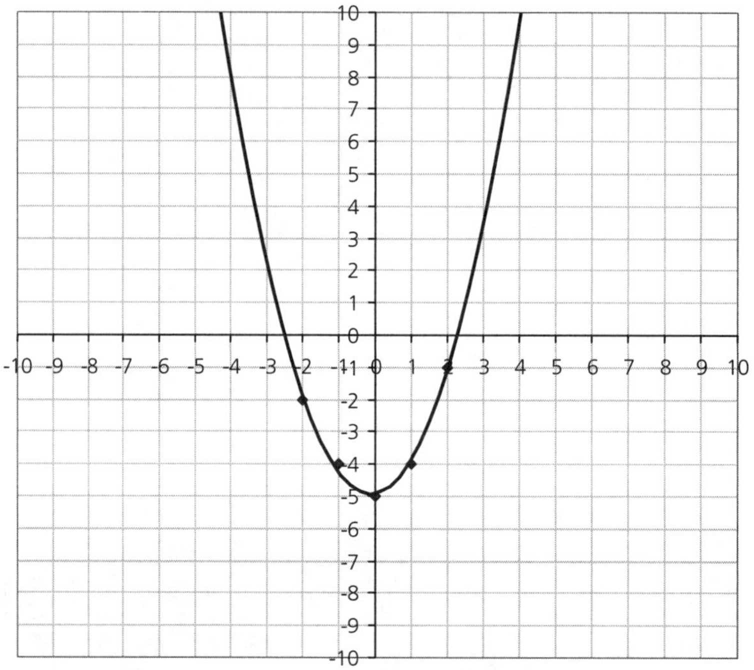

x-intercepts: ***$(\sqrt{5}, 0)$; $(-\sqrt{5}, 0)$***

y-intercept: ***(0, -5)***

Assignment #1 for Finding Zeros by Factoring

Name: ______________________________

1. Complete each of the following multiplication tables to factor the quadratic function and to find its *x*-intercepts.

   a. $y = x^2 + x - 12 = ($ ***x + 4*** $)($ ***x - 3*** $)$

   x-intercepts: ***(-4,0) and (3,0)***

•	x	***4***
x	x^2	***4x***
-3	***-3x***	***-12***

   b. $y = x^2 - 8x + 7 = ($ ***x - 7*** $)($ ***x - 1*** $)$

   x-intercepts: ***(7,0) and (1,0)***

•	x	***-7***
x	x^2	***-7x***
-1	***-x***	***7***

   c. $y = x^2 + 9x - 90 = ($ ***x + 15*** $)($ ***x - 6*** $)$

   x-intercepts: ***(-15,0) and (6,0)***

•	x	***15***
x	x^2	***15x***
-6	***-6x***	***-90***

2. Find the x-intercepts and the factors of each of the following quadratic functions.

   a. $y = x^2 - 10x + 24 = ($ ***x - 6*** $)($ ***x - 4*** $)$

   x-intercepts: ***(6,0) and (4,0)***

   b. $y = x^2 - 2x - 63 = ($ ***x − 9*** $)($ ***x + 7*** $)$

   x-intercepts: ***(9,0) and (-7,0)***

   c. $y = x^2 + 2x - 3 = ($ ***x + 3*** $)($ ***x - 1*** $)$

   x-intercepts: ***(-3,0) and (1,0)***

3. Find the roots (solutions) of each of the quadratic equations.

   a. $x^2 - 10x - 56 = 0$

      roots: ***14 and -4***

   b. $x^2 + 18x + 32 = 0$

      roots: ***-16 and -2***

   c. $x^2 + 11x - 60 = 0$

      roots: ***-15 and 4***

Assignment #2 for Finding Zeros by Factoring

Name: ______________________________

Factor each of the following quadratic functions and find their *x*-intercepts.

1. $y = x^2 + 5x + 6 = ($ ***x + 2*** $)($ ***x + 3*** $)$ *x*-intercepts: ***(-2,0) (-3,0)***

2. $y = x^2 - 5x + 6 = ($ ***x – 2*** $)($ ***x - 3*** $)$ *x*-intercepts: ***(2,0) (3,0)***

3. $y = x^2 + x - 6 = ($ ***x + 3*** $)($ ***x - 2*** $)$ *x*-intercepts: ***(-3,0) (2,0)***

4. $y = x^2 - 5x - 6 = ($ ***x - 6*** $)($ ***x + 1*** $)$ *x*-intercepts: ***(6,0) (-1,0)***

5. $y = x^2 - 7x + 6 = ($ ***x - 6*** $)($ ***x – 1*** $)$ *x*-intercepts: ***(6,0) (1,0)***

6. $y = x^2 + 7x + 6 = ($ ***x + 6*** $)($ ***x + 1*** $)$ *x*-intercepts: ***(-6,0) (-1,0)***

7. $y = x^2 + 5x - 6 = ($ ***x + 6*** $)($ ***x - 1*** $)$ *x*-intercepts: ***(-6,0) (1,0)***

8. $y = x^2 - 5x - 6 = ($ ***x - 6*** $)($ ***x + 1*** $)$ *x*-intercepts: ***(6,0) (-1,0)***

9. $y = x^2 + 5x + 4 = ($ ***x + 4*** $)($ ***x + 1*** $)$ *x*-intercepts: ***(-4,0) (-1,0)***

10. $y = x^2 - 5x - 24 = ($ ***x - 8*** $)($ ***x + 3*** $)$ *x*-intercepts: ***(8,0) (-3,0)***

11. $y = x^2 + 5x - 36 = ($ ***x + 9*** $)($ ***x - 4*** $)$ *x*-intercepts: ***(-9,0) (4,0)***

12. $y = x^2 + 4x - 12 = ($ ***x + 6*** $)($ ***x - 2*** $)$ *x*-intercepts: ***(-6,0) (2,0)***

13. $y = x^2 - 3x + 2 = ($ ***x - 2*** $)($ ***x - 1*** $)$ *x*-intercepts: ***(2,1) (1,0)***

14. $y = x^2 - x - 56 = ($ ***x - 8*** $)($ ***x + 7*** $)$ *x*-intercepts: ***(8,0) (-7,0)***

15. $y = x^2 - x - 42 = (\ \mathbf{x - 7}\ )(\ \mathbf{x + 6}\ )$ x-intercepts: ***(7,0) (-6,0)***

16. $y = x^2 + x - 20 = (\ \mathbf{x + 5}\ )(\ \mathbf{x - 4}\ )$ x-intercepts: ***(-5,0) (4,0)***

17. $y = x^2 - 25 = (\ \mathbf{x + 5}\ )(\ \mathbf{x - 5}\ )$ x-intercepts: ***(-5,0) (5,0)***

18. $y = x^2 - 6x = (\ \mathbf{x}\ )(\ \mathbf{x - 6}\ )$ x-intercepts: ***(0,0) (6,0)***

19. $y = x^2 - 16 = (\ \mathbf{x + 4}\ )(\ \mathbf{x - 4}\ )$ x-intercepts: ***(-4,0) (4,0)***

20. $y = x^2 - 7x - 18 = (\ \mathbf{x - 9}\ )(\ \mathbf{x + 2}\ )$ x-intercepts: ***(9,0) (-2,0)***

21. $y = x^2 - 4x - 12 = (\ \mathbf{x - 6}\ )(\ \mathbf{x + 2}\ )$ x-intercepts: ***(6,0) (-2,0)***

22. $y = x^2 - 6x + 5 = (\ \mathbf{x - 5}\ )(\ \mathbf{x - 1}\ )$ x-intercepts: ***(5,0) (1,0)***

23. $y = x^2 - 2x + 1 = (\ \mathbf{x - 1}\ )(\ \mathbf{x - 1}\ )$ x-intercepts: ***(1,0)***

24. $y = x^2 - 8x + 7 = (\ \mathbf{x - 7}\ )(\ \mathbf{x - 1}\ )$ x-intercepts: ***(7,0) (1,0)***

25. $y = x^2 - 4x - 165 = (\ \mathbf{x - 15}\ )(\ \mathbf{x + 11}\ )$ x-intercepts: ***(15,0) (-11,0)***

26. $y = x^2 + 18x - 63 = (\ \mathbf{x + 21}\ )(\ \mathbf{x - 3}\ )$ x-intercepts: ***(-21,0) (3,0)***

27. $y = x^2 - 11x + 10 = (\ \mathbf{x - 10}\ )(\ \mathbf{x - 1}\ )$ x-intercepts: ***(10,0) (1,0)***

28. $y = x^2 + 6x - 55 = (\ \mathbf{x + 11}\ )(\ \mathbf{x - 5}\ )$ x-intercepts: ***(-11,0) (5,0)***

29. $y = x^2 - 9x + 8 =$ (***x - 8***)(***x - 1***) x-intercepts: ***(8,0) (1,0)***

30. $y = x^2 - 13x + 12 =$ (***x - 12***)(***x - 1***) x-intercepts: ***(12,0) (1,0)***

31. $y = x^2 - 13x =$ (***x***)(***x - 13***) x-intercepts: ***(0,0) (13,0)***

32. $y = x^2 - 12x + 36 =$ (***x - 6***)(***x - 6***) x-intercepts: ***(6,0)***

33. $y = x^2 - 14x + 49 =$ (***x - 7***)(***x - 7***) x-intercepts: ***(7,0)***

34. $y = x^2 - 625 =$ (***x + 25***)(***x - 25***) x-intercepts: ***(-25,0) (25,0)***

35. $y = x^2 - 196 =$ (***x - 14***)(***x + 14***) x-intercepts: ***(14,0) (-14,0)***

36. $y = x^2 - 16x + 64 =$ (***x - 8***)(***x - 8***) x-intercepts: ***(8,0)***

37. $y = x^2 - 5x =$ (***x***)(***x - 5***) x-intercepts: ***(0,0) (5,0)***

38. $y = x^2 - 1x =$ (***x***)(***x - 1***) x-intercepts: ***(0,0) (1,0)***

39. $y = x^2 + 18x + 81 =$ (***x + 9***)(***x + 9***) x-intercepts: ***(-9,0)***

40. $y = x^2 + 20x + 100 =$ (***x + 10***)(***x + 10***) x-intercepts: ***(-10,0)***

Assignment #1 for Finding Zeros using the Quadratic Formula

Name: ______________________________

General form of the quadratic equation:

$$y = Ax^2 + Bx + C$$

In this equation, *A*, *B*, and *C* are real numbers, but *A* is not zero.

Quadratic formula:

$$x = \frac{-B \pm \sqrt{B^2 - 4AC}}{2A}$$

For each quadratic equation:

- Write the values of *A*, *B*, and *C* from the general form
- Use the quadratic formula to solve the equation for the roots (and the *x*-intercepts)
- Use the roots (or the *x*-intercepts) to write the equation in factored form

$y = x^2 - 5x - 14 = ($ ***x - 7*** $)($ ***x + 2*** $)$

$A =$ **1** $\quad B =$ **-5** $\quad C =$ **-14**

Solve the equation:

$$\boldsymbol{x} = \frac{\boldsymbol{5 \pm \sqrt{25 - 4(1)(-14)}}}{\boldsymbol{2(1)}}$$

$$\boldsymbol{x} = \frac{\boldsymbol{5 \pm \sqrt{81}}}{\boldsymbol{2}}$$

$$\boldsymbol{x} = \frac{\boldsymbol{5 \pm 9}}{\boldsymbol{2}}$$

x = 7 or x = -2

x-intercepts: ***(7,0) (-2,0)***

$y = x^2 - 12x - 13 = ($ ***x - 13*** $)($ ***x + 1*** $)$

$A =$ ***1*** $B =$ ***-12*** $C =$ ***-13***

Solve the equation:

$$\boldsymbol{x = \frac{12 \pm \sqrt{144 - 4(1)(-13)}}{2(1)}}$$

$$\boldsymbol{x = \frac{12 \pm \sqrt{196}}{2}}$$

$$\boldsymbol{x = \frac{12 \pm 14}{2}}$$

x = 13 or x = -1

x-intercepts: ***(13,0) (-1,0)***

$y = x^2 + 8x - 33 = ($ ***x + 11*** $)($ ***x - 3*** $)$

$A =$ ***1*** $B =$ ***8*** $C =$ ***-33***

Solve the equation:

$$\boldsymbol{x = \frac{-8 \pm \sqrt{64 - 4(1)(-33)}}{2(1)}}$$

$$\boldsymbol{x = \frac{-8 \pm \sqrt{196}}{2}}$$

$$\boldsymbol{x = \frac{-8 \pm 14}{2}}$$

x = 3 or x = -11

x-intercepts: ***(3,0) (-11,0)***

Assignment #2 for Finding Zeros using the Quadratic Formula

Name: ____________________

General form of the quadratic equation:

$$y = Ax^2 + Bx + C$$

In this equation, A, B, and C are real numbers, but A is not zero.

Quadratic formula:

$$x = \frac{-B \pm \sqrt{B^2 - 4AC}}{2A}$$

For each quadratic equation:

- Write the values of A, B, and C from the general form
- Use the quadratic formula to solve the equation for the roots (and the x-intercepts)
- Find the y-intercept
- Find the vertex
- Use the average of the x-intercepts to find the equation of the axis of symmetry
- Calculate the sum of the roots and the product of the roots
- Use the roots (or the x-intercepts) to write the equation in factored form
- Sketch the graph

$y = x^2 - 6x + 8 = ($ x - 4 $)($ x - 2 $)$

$A =$ **1** $B =$ **-6** $C =$ **8**

Solve the equation:

$$x = \frac{6 \pm \sqrt{36 - 4(1)(8)}}{2(1)}$$

$$x = \frac{6 \pm 2}{2}$$

$$x = \frac{6 \pm \sqrt{4}}{2}$$

x = 4 or x = 2

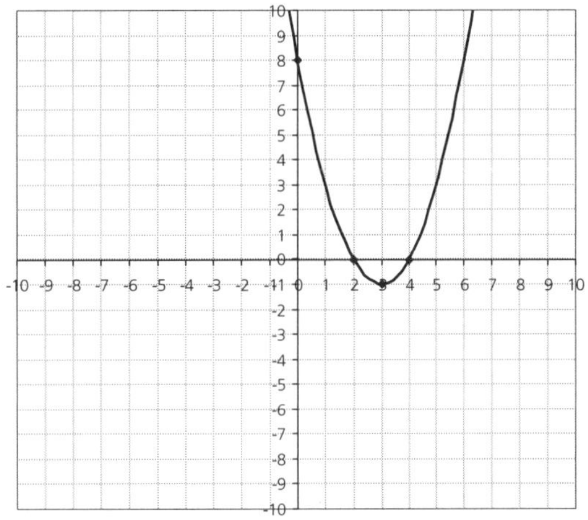

x-intercepts: **(4,0) (2,0)**

y-intercept: **(0,8)**

Axis of symmetry: $x =$ **3**

Vertex: (**3** , **-1**)

Sum of roots = **6**

Product of roots = **8**

$y = x^2 - 8x + 12 = (\ \boldsymbol{x - 6}\ )(\ \boldsymbol{x - 2}\ )$

$A =$ **1** $\quad B =$ **-8** $\quad C =$ **12**

Solve the equation:

$$\boldsymbol{x} = \frac{\boldsymbol{8 \pm \sqrt{64 - 4(1)(12)}}}{\boldsymbol{2(1)}} \qquad \boldsymbol{x} = \frac{\boldsymbol{8 \pm 4}}{\boldsymbol{2}}$$

$$\boldsymbol{x} = \frac{\boldsymbol{8 \pm \sqrt{16}}}{\boldsymbol{2}} \qquad \boldsymbol{x = 6 \text{ or } x = 2}$$

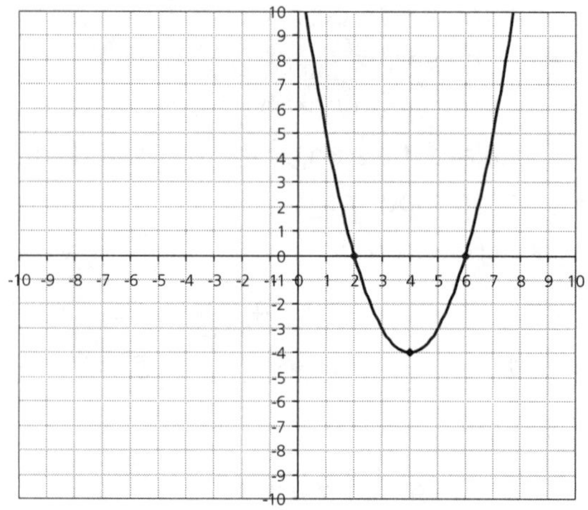

x-intercepts: ***(6,0) (2,0)***

y-intercept: ***(0,12)***

Axis of symmetry: $x =$ ***4***

Vertex: (***4*** , ***-4***)

Sum of roots = ***8***

Product of roots = ***12***

Assignment for Modeling: Vertical Motion

Name: ______________________________

Mark McGwire hits a foul ball straight up from a height of four feet off the ground. The following equation models the vertical motion of the ball:

$$y = -16t^2 + vt + h$$

In this equation, y is the number of feet the ball is above the ground, t is the time in seconds since the ball was hit, v is the initial velocity (or speed) of the ball in feet per second, and h is the height at which the ball was originally hit.

1. If the initial velocity of the foul ball is 130 feet per second, write the equation for the foul ball.

   Equation: $\underline{-16t^2 + 130t + 4}$

Use the equation from problem 1 to answer the following questions.

2. How high is the ball after:

   1 second after it is hit?

   $y = -16(1)^2 + 130(1) + 4 = -16 + 130 + 4 = 118$
   The ball is 118 feet above the ground 1 second after it is hit.

   2 seconds?

   $y = -16(2)^2 + 130(2) + 4 = -64 + 260 + 4 = 200$
   The ball is 200 feet above the ground 2 seconds after it is hit.

   10 seconds?

   $y = -16(10)^2 + 130(10) + 4 = -1600 + 1300 + 4 = -296$
   The ball is -296 feet above the ground 10 seconds after it is hit.

3. Do all of the answers to question 2 make sense in this problem situation?

   No. The height after 10 seconds does not make sense, because the ball cannot go below ground by 296 feet.

4. Use the quadratic formula to find when the ball is on the ground. How do you know?

   $$x = \frac{-130 \pm \sqrt{130^2 - 4(-16)(4)}}{2(-16)} \qquad x = \frac{-130 \pm 130.98}{-32}$$

   $$x = \frac{-130 \pm \sqrt{17156}}{-32} \qquad x = -0.03 \text{ or } x = 8.16$$

   The ball will hit the ground after about 8.16 seconds because that is when the height is 0 feet.

5. When is the ball at its highest point? How do you know? What is its height at this point?

   The ball will be at a maximum height of 286.06 feet after about 4.06 seconds because the vertex of the curve is at (4.06, 286.06).

6. Use the quadratic formula to find when the ball will be back at the level from which it was hit.

   4 = -16t² + 130t + 4, solving for 0, 0 = -16t² + 130t

   $$x = \frac{-130 \pm \sqrt{130^2 - 4(-16)(0)}}{2(-16)} \qquad x = \frac{-130 \pm 130}{-32}$$

   $$x = \frac{-130 \pm \sqrt{16900}}{-32} \qquad x = 0 \text{ or } x = 8.125$$

   The ball will be back at 4 feet above the ground after 8.125 seconds.

7. When is the ball 200 feet off the ground?
   (Hint: Think about whether there is more than one answer.)

   Using an answer from question 2, it is already known that the ball is 200 feet off the ground after 2 seconds. The parabola is symmetric, so there is a second time when the ball is 200 feet off the ground. The vertex is the highest point, which is reached after about 4.06 seconds. Therefore, the ball will be at 200 feet 6.12 seconds after it is hit. (2 is 2.06 to the left of 4.06 and 6.12 is 2.06 to the right of 4.06)

8. Using the information you found for this problem situation, complete the following chart.

Time		Height Above Ground
1 second		***118 feet***
2 seconds		***200 feet***
10 seconds		***-296 feet***
4.06 sec		***268.06 feet***
-0.03 sec	***8.16 sec***	0 feet (on the ground)
0 sec	***8.125 sec***	4 feet
2 sec	***6.12 sec***	200 feet

9. Graph the equation for this problem situation.

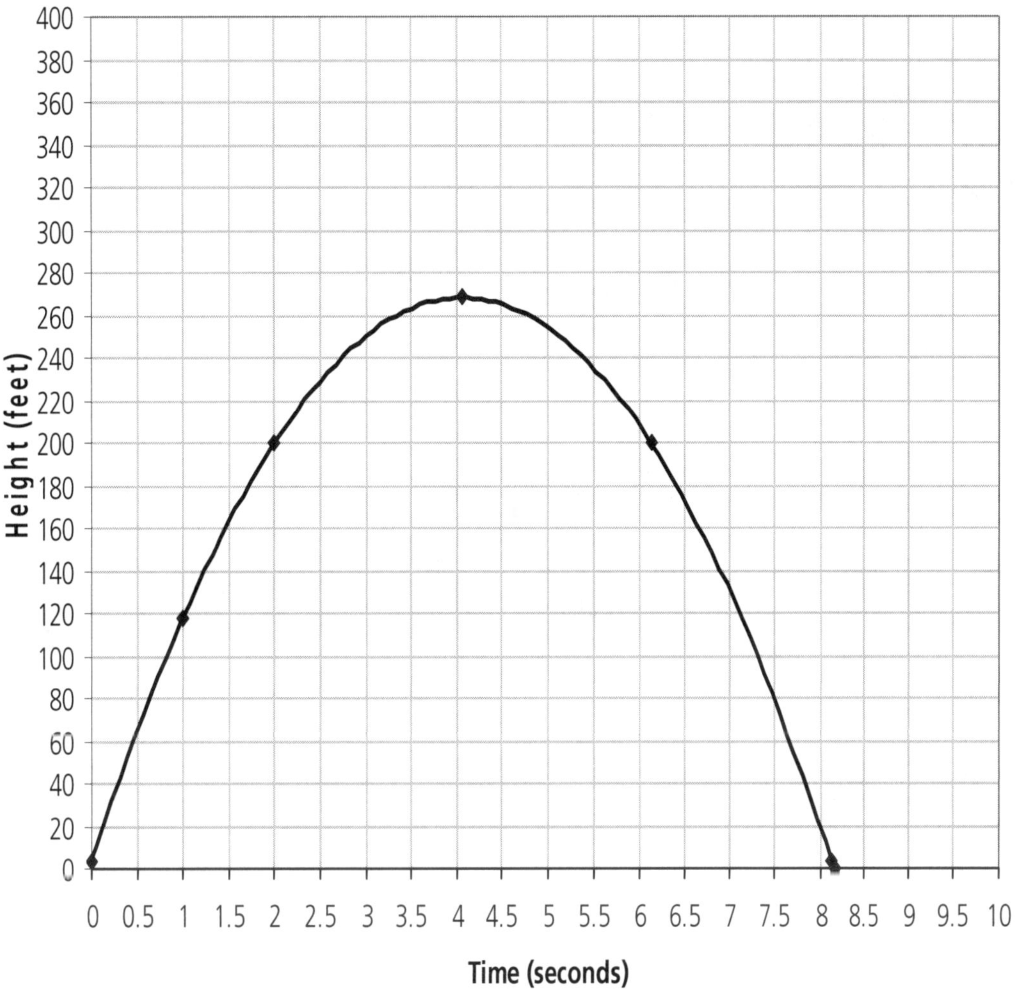

Assignment for Measures of Central Tendency and Random Numbers

Name: ______________________________

Calculate the mean, the median, and the mode for each set of data.

1. Data: 15, 16, 19, 20, 20, 21, 22

   mean: $\bar{x} = \frac{\Sigma x}{n}$ = ***19*** median = ***20*** mode = ***20***

   mean = 133/7

2. Data: 111, 111, 313, 415, 612

   mean: $\bar{x} = \frac{\Sigma x}{n}$ = ***312.4*** median = ***313*** mode = ***111***

   mean = 1562/5

3. Data: 11, 54, 65, 98, 201

   mean: $\bar{x} = \frac{\Sigma x}{n}$ = ***85.8*** median = ***65*** mode = ***no mode***

   mean = 429/5

4. Data: 24, 26, 28, 30, 30, 32

   mean: $\bar{x} = \frac{\Sigma x}{n}$ = ***28.33*** median = ***29*** mode = ***30***

   mean = 170/6 ≈ 28.33 ***median = 29***

Assignment for Baseball History and Compact Discs

Name ______________________________

Give several examples of when the median value is a more appropriate statistic to use than the mean.

Answers will vary. Some appropriate answers include home selling prices, salaries, expenditures, expenses and other situations that have outliers that would affect the mean.

Give several examples of when the mean and the median provide the same information.

Answers will vary. Test scores might be an example.

Give an example of where the means are the same, but the standard deviations are different.

Answers will vary.

Assignment for Breakfast Cereals and Airline Industry

Name ______________________________

When the results of standardized tests are reported you will often find the scores represented in by quartile. This tells you your standing with respect to others who took the test.

Let's look at the mean scores for males and females on the verbal and mathematics SAT's from 1967 – 2000.

Average SAT® scores of entering college classes, 1967-2000

	Male		Female		All	
Year	Verbal	Math	Verbal	Math	Verbal	Math
1967	540	535	545	495	543	516
1968	541	533	543	497	543	516
1969	536	534	543	498	540	517
1970	536	531	538	493	537	512
1971	531	529	534	494	532	513
1972	531	527	529	489	530	509
1973	523	525	521	489	523	506
1974	524	524	520	488	521	505
1975	515	518	509	479	512	498
1976	511	520	508	475	509	497
1977	509	520	505	474	507	496
1978	511	517	503	474	507	494
1979	509	516	501	473	505	493
1980	506	515	498	473	502	492
1981	508	516	496	473	502	492
1982	509	516	499	473	504	493
1983	508	516	498	474	503	494
1984	511	518	498	478	504	497
1985	514	522	503	480	509	500
1986	515	523	504	479	509	500
1987	512	523	502	481	507	501
1988	512	521	499	483	505	501

	Male		Female		All	
Year	Verbal	Math	Verbal	Math	Verbal	Math
1989	510	523	498	482	504	502
1990	505	521	496	483	500	501
1991	503	520	495	482	499	500
1992	504	521	496	484	500	501
1993	504	524	497	484	500	503
1994	501	523	497	487	499	504
1995	505	525	502	490	504	506
1996	507	527	503	492	505	508
1997	507	530	503	494	505	511
1998	509	531	502	496	505	512
1999	509	531	502	495	505	511
2000	507	533	504	498	505	514

Source: www.collageboard.org

1. Represent the data using a visual display. Discuss why you chose the specific display you did.

   Answers will vary.

To answer the questions, you will need to define variables and variable type. To make comparisons, you may make observations, but you will want to back your observation with specific statistics. Name the statistics you think you will want to use to answer the questions below.

2. Compare the data on males and females on

   verbal portion of the SAT

   Answers will vary but could include them mean, standard deviation and five number summary.

   math portion of the SAT

   Answers will vary but could include the mean, standard deviation and five number summary.

   total means

   Answers will vary but could include the mean, standard deviation and five number summary.

3. Compare the scores of males between verbal and math.

   Answers will vary but will probably state that males do better on the math section.

4. Compare the scores of females between verbal and math

   Answers will vary but will probably state that females do better on the verbal section.

5. Discuss any trends you find in the data.

   Answers will vary but should show that from 1967 to 2000, males' scores on the verbal section have gone down. Math scores went down for about 30 years and are going up and almost equal to the 1967 score. For females, scores on the verbal section went downward with a slight upward trend; however, they never regained the high mean in 1967. Math scores have been fairly stable with only a slight dip in the 1970's and 1980's. The overall trends tend to match the trends of the females more than males for verbal and math.

6. What are the standard deviations for scores for males on verbal, on math and total and for females on verbal, math and total?

   Standard deviations:

	Verbal	***Math***	***Total***
Male	***11.34***	***5.90***	***15.53***
Female:	***15.45***	***8.48***	***21.52***

7. Is there anything you understand better about the relationship of these scores after computing the standard deviations? Do the standard deviations differ? How do they compare. If there are differences, in which data groups do you see these differences?

   Answers will vary.

   Statistics from the data set that the students may use:

	Verbal		*Math*		*Total*	
	Male	*Female*	*Male*	*Female*	*Male*	*Female*
MEAN	*514.21*	*508.56*	*523.76*	*484.68*	*1037.97*	*993.24*
MINIMUM	*501*	*495*	*515*	*473*	*1021*	*969*
Q_1	*507*	*498*	*520*	*478*	*1026*	*979*
MEDIAN	*509.5*	*502.5*	*523*	*483.5*	*1033*	*983*
Q_3	*515*	*509*	*529*	*493*	*1040*	*1002*
MAXIMUM	*541*	*545*	*535*	*498*	*1075*	*1041*

Assignment for Variance and Candy Sale

Name ______________________________

Hikers around America are feeling short changed. There is a belief that there are not enough hiking trails and that the number of miles available is limited. Collect some data that either supports or negates this claim. Your analysis should include a data table and statistical evidence along with a summary of your findings. To help you in the process, you might try looking at the following website: www.TrailLink.com.

Students should create a visual display (or displays) that will most appropriately show the number of trails and miles of trails available.

Students should also be deciding between stem and leaf plots and histograms as choices for their visual displays.

There should be a discussion regarding the distribution as a means of justifying the claim that there are not enough miles.

Assignment for College Students and Retail Sales

Name ______________________________

You have a hypothesis that the cost of a box of cereal is a function of the grams of sugar per serving. You believe the more sugar, the higher the cost. To determine whether your hypothesis is correct, go to the nearest market and select 15 different cereals. Set up an appropriate analysis of the data and explain your findings. Be sure to note your sampling technique and to include a visual display of the data in addition to the table.

Students should write the hypothesis in their own words. They should also indicate their sampling technique, which should be random.

Students should choose a display that best determines if one factor is a function of the other. For example, students might choose a line graph or a number line plot.

Assignment for Stroop Test

Name: ______________________________

On this grid, use the data from your table for the matching lists to construct a graph with the length of the list on the horizontal axis and the average time on the vertical axis. Label the axes on your graph.

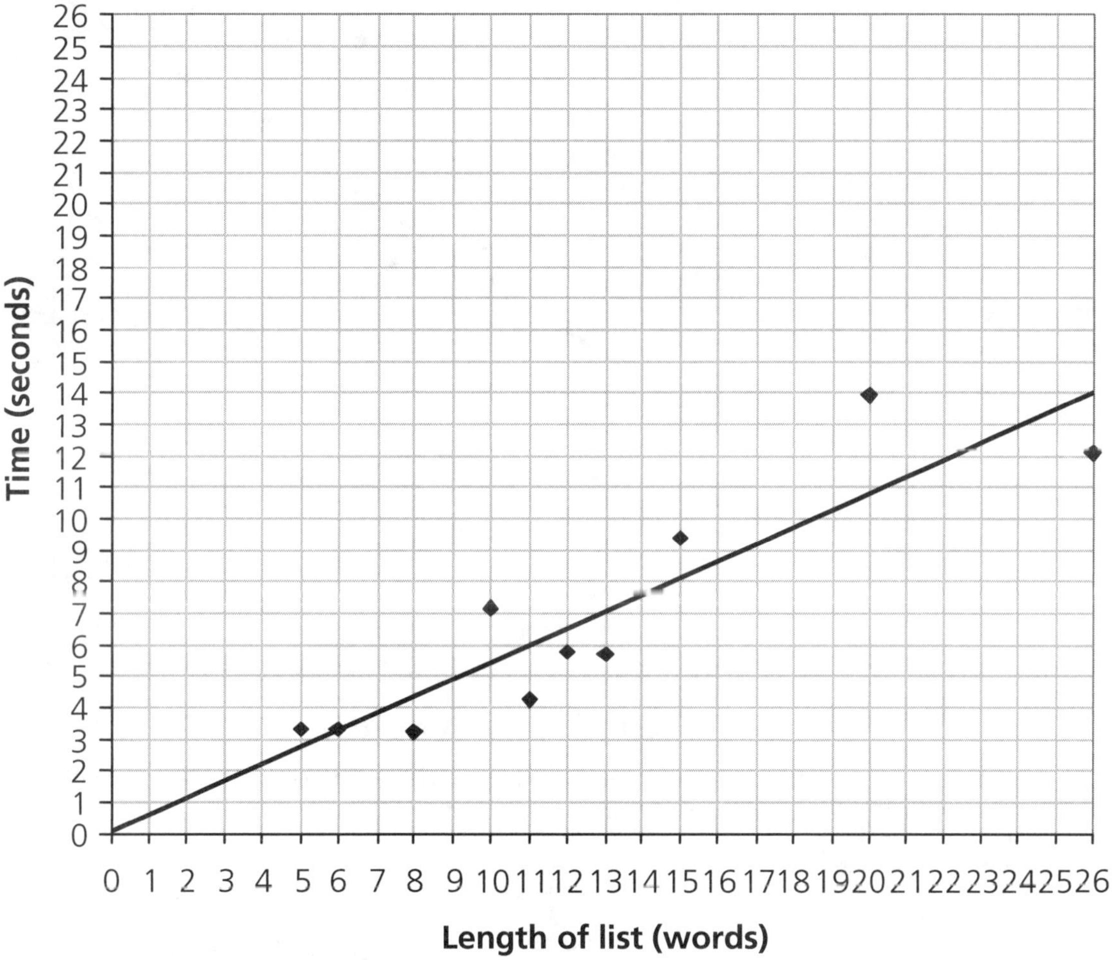

These graphs are based upon the sample data given in the Teacher Text. Your class' graphs will vary based upon the actual data collected in your class. However, axes labels should be as indicated.

On this grid, use the data from your table for the non-matching lists to construct a graph with the length of the list on the horizontal axis and the average time on the vertical axis. Label the axes on your graph.

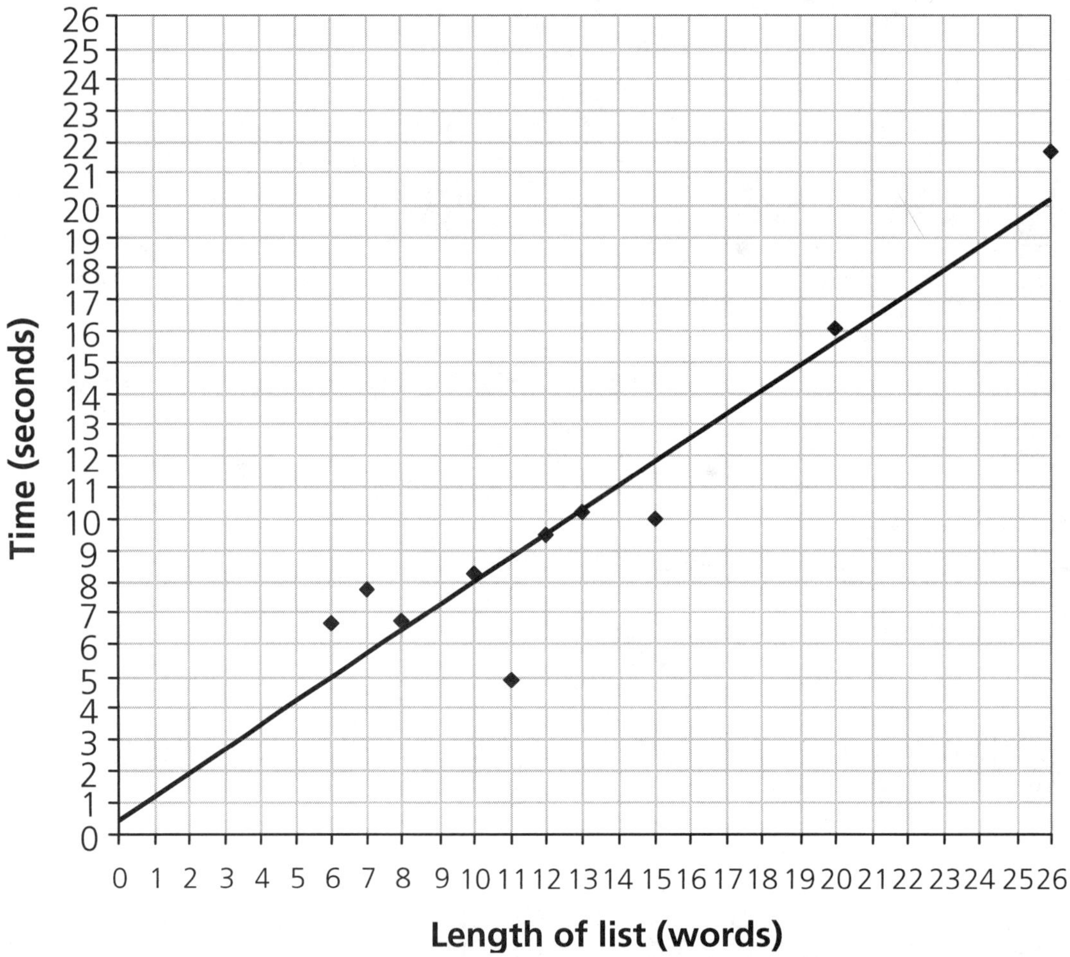

Use the equations of the line of best fit for matching lists and the line of best fit for the non-matching lists to answer the following questions.

1. How long would it take to read a list of:

    a. 100 words on a matching list?

    b. 100 words on a non-matching list?

    c. 1000 words on a matching list?

    d. 1000 words on a non-matching list?

For each, use the two regression equations found for the matching and non-matching lists. Substitute given independent variable values for x, and solve the dependent variable values.

Use the equations of the linear regression lines for the matching lists and the non-matching lists to answer the following questions.

2. How many matching words could a person read in:
   a. 10 minutes?

   b. 1 hour?

For each, use the two regression equations found for the matching and non-matching lists. Substitute given dependent variable values for y, and solve the independent variable values.

3. How many *non-matching* words could a person read in:
   a. 10 minutes?

   b. 1 hour?

4. Do you think these answers are reasonable? Explain your answer.

   Answers will vary.

   The further the given value is from the actual data collected, the less reliable it will be.

   Students may see the answers as reasonable for the requested number of words or length of list. However, it may be unreasonable to read a very long list or to continue reading for the requested amount of time.

5. What do you think accounts for the differences in the times and the number of words for the two different lists?

   Answers will vary.

   In general, times for matching lists should be less than comparable times for non-matching lists. Differences may be accounted for, in part, by the conflicting pieces of information being processed when a non-matching list is presented.

Assignment for Hanford Nuclear Power Plant

Name: ______________________________

Use the equation of the linear regression line for the *Hanford Nuclear Power Plant* data to answer the following questions. $\boldsymbol{y = 9.28x + 114.7}$

1. What is the predicted cancer rate per 100,000 residents of a community where the exposure index is:

   a. 10.5? $y = 9.28(10.5) + 114.7$

   $\boldsymbol{y = 97.44 + 114.7}$

   $\boldsymbol{y = 212.14}$ ***The expected cancer rate is 212 people per 100,000 residents.***

   b. 21.7? $y = 9.28(21.7) + 114.7$

   $\boldsymbol{y = 201.376 + 114.7}$

   $\boldsymbol{y = 316.076}$ ***The cancer rate is 316 per 100,000 residents.***

2. What is the predicted exposure index of a community that has a cancer rate of:

   a. 1,000 per 100,000 residents? $\boldsymbol{1000 = 9.28x + 114.7}$

   $\boldsymbol{885.3 - 9.28x}$

   $\boldsymbol{x = 95.40}$

   The predicted exposure index is 95.40.

   b. 1,500 per 100,000 residents? $\boldsymbol{1500 = 9.28x + 114.7}$

   $\boldsymbol{1385.3 = 9.28x}$

   $\boldsymbol{x = 149.28}$

   The predicted exposure index is 149.28.

Assignment for Human Chain: Wrist Experiment

Name: ____________________________

Use the data in this table to solve problem 1 on the next page.

State Expenditures per Student in Public Education 1996-97					
Rank	State	Expenditures per Student ($/student)	Rank	State	Expenditures per Student ($/student)
51	UT	3,837	25	GA	5,585
50	AZ	4,048	24	HI	5,720
49	AR	4,172	23	IA	5,720
48	OK	4,187	22	WA	5,805
47	MS	4,269	21	IN	5,886
46	ID	4,500	20	VA	5,920
45	LA	4,527	19	OR	5,988
44	ND	4,667	18	NH	6,014
43	SD	4,680	17	WY	6,036
42	TN	4,898	16	MN	6,041
41	NM	4,927	15	ME	6,385
40	MI	4,949	14	WV	6,406
39	NV	4,998	13	VT	6,503
38	NC	5,028	12	WI	6,521
37	SC	5,105	11	MD	6,547
36	CO	5,147	10	MI	6,654
35	NE	5,250	9	PA	6,967
34	AL	5,255	8	MA	7,069
33	CA	5,284	7	DE	7,086
32	KY	5,346	6	D.C.	7,175
31	MT	5,380	5	RI	7,665
30	IL	5,423	4	CT	8,376
29	FL	5,427	3	NY	8,658
28	KS	5,493	2	AK	8,900
27	OH	5,527	1	NJ	9,455
26	TX	5,551			

Source: *The New York Times 2000 Almanac*, p. 370

1. Construct a stem-and-leaf plot for the data on state expenditures in public education.

3	4	5	6	7	8	9
		988				
		920				
		886				
		805				
		720				
		720				
		585				
		551				
	998	***527***				
	949	***493***				
	927	***427***	***967***			
	898	***423***	***654***			
	680	***380***	***547***			
	667	***346***	***521***			
	527	***284***	***503***			
	500	***255***	***406***			
	269	***250***	***385***	***665***		
	187	***147***	***041***	***175***	***900***	
	172	***105***	***036***	***086***	***658***	
837	***048***	***028***	***014***	***069***	***376***	***455***

State Expenditures per Student

Stem = $1000s

2. Use the stem-and-leaf plot (and look at the data table) to describe the distribution of the data. Write a report that describes the distribution of the data in terms of clusters of values, overall pattern or trend to the data, and outliers (values that differ greatly from the others).

Answers will vary, but should include the following observations:

The distribution is skewed to the right, because the majority of values are smaller.

The majority of values cluster between $4000 and $6000.

$3837 and $9455 are the extreme values, but are not technically outliers.

Assignment for Human Chain: Shoulder Experiment

Name: ____________________________

1. Prepare a five-number summary, as well as calculating the mean for the data in the table below.

Number of Representatives, by State (1997)					
Rank	State	Number of Reps	Rank	State	Number of Reps
1	CA	52	27	CT	6
2	NY	31	28	OK	6
3	TX	30	29	OR	5
4	FL	23	30	IA	5
5	PA	21	31	MS	5
6	IL	20	32	KS	4
7	OH	19	33	AR	4
8	MI	16	34	WV	3
9	NJ	13	35	UT	3
10	NC	12	36	NE	3
11	GA	11	37	NM	3
12	VA	11	38	ME	2
13	MA	10	39	NV	2
14	IN	10	40	NH	2
15	MO	9	41	HI	2
16	WI	9	42	ID	2
17	TN	9	43	RI	2
18	WA	9	44	MT	1
19	MD	8	45	SD	1
20	MN	8	46	DE	1
21	LA	7	47	ND	1
22	AL	7	48	D.C.	--
23	KY	6	49	VT	1
24	AZ	6	50	AK	1
25	SC	6	51	WY	1
26	CO	6	Total	US	435

Five-Number Summary

Low: ***1***

1st Quartile: ***2***

Median: ***6***

3rd Quartile: ***10***

High: ***52***

Mean: ***8.53 based on 51 entities or 8.7 based on 50 states***

2. Construct a box-and-whisker plot for the data.

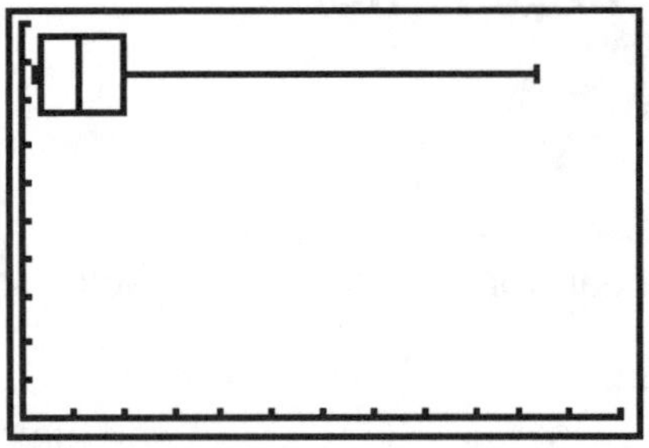

0 10 20 30 40 50

3. Write a summary report about the data.

   Answers will vary but should include a discussion of outliers, clusters, distributions, etc.

Assessments

Contents

Assessments

Unit 1: Organizing Single Variable Data

Unit 2: Patterns and Linear Functions

Unit 3: Proportional Reasoning

Unit 4: Modeling Situations Using Multiple Representations

Unit 5: Linear Equations: Solving, Graphing, Intercepts, and Slope

Unit 6: Area and Perimeter

Unit 7: The Geometry and Algebra of Right Triangles

Unit 8: Quadratics

Unit 9: Statistical Analysis

Unit 10: Probability

Baseball

Name: ______________________

The data set below represents the top 40 homerun hitters in the American and National Baseball Leagues for 2001.

American League Homerun Leaders 2001	National League Homerun Leaders 2001
1) Alex Rodriguez, Tex 52	1) Barry Bonds, SF 73
2) Jim Thome, Cle 49	2) Sammy Sosa, ChC 64
3) R. Palmeiro, Tex 47	3) Luis Gonzalez, Ari 57
4) Manny Ramirez, Bos 41	4) Shawn Green, LA 49
5) Troy Glaus, Ana 41	5) Todd Helton, Col 49
6) Carlos Delgado, Tor 39	6) Richie Sexson, Mil 45
7) Jason Giambi, Oak 38	7) Phil Nevin, SD 41
8) Bret Boone, Sea 37	8) Jeff Bagwell, Hou 39
9) Juan Gonzalez, Cle 35	9) Chipper Jones, Atl 38
10) Jose Cruz, Tor 34	10) Larry Walker, Col 38
11) Tino Martinez, NYY 34	11) Rich Aurilia, SF 37
12) Paul Konerko, CWS 32	12) Brian Giles, Pit 37
13) Eric Chavez, Oak 32	13) Albert Pujols, StL 37
14) Miguel Tejada, Oak 31	14) Mike Piazza, NYM 36
15) M. Ordonez, CWS 31	15) Gary Sheffield, LA 36
16) Mike Sweeney, KC 29	16) Jeromy Burnitz, Mil 34
17) Jose Valentin, CWS 28	17) Andruw Jones, Atl 34
18) G. Anderson, Ana 28	18) V. Guerrero, Mon 34
19) Ellis Burks, Cle 28	19) Aramis Ramirez, Pit 34
20) Raul Mondesi, Tor 27	20) Lance Berkman, Hou 34
21) Trot Nixon, Bos 27	21) Reggie Sanders, Ari 33
22) Torii Hunter, Min 27	22) Cliff Floyd, Fla 31
23) Jermaine Dye, KC/Oak 26	23) Bobby Abreu, Phi 31
24) Corey Koskie, Min 26	24) Ryan Klesko, SD 30
25) B. Williams, NYY 26	25) Jim Edmonds, StL 30
26) Mike Cameron, Sea 25	26) Mark McGwire, StL 29
27) Tony Batista, Bal/Tor 25	27) J.D. Drew, StL 27
28) Ivan Rodriguez, Tex 25	28) Pat Burrell, Phi 27
29) Carlos Beltran, KC 24	29) Moises Alou, Hou 27
30) Carlos Lee, CWS 24	30) Jose Hernandez, Mil 25
31) Greg Vaughn, TB 24	31) Brian Jordan, Atl 25
32) Ruben Sierra, Tex 23	32) Scott Rolen, Phi 25
33) Edgar Martinez, Sea 23	33) Bubba Trammell, SD 25
34) Jorge Posada, NYY 22	34) Paul Lo Duca, LA 25
35) Brian Daubach, Bos 22	35) Lee Stevens, Mon 25
36) Derek Jeter, NYY 21	36) Vinny Castilla, Hou 23
37) Paul O'Neill, NYY 21	37) Preston Wilson, Fla 23
38) John Olerud, Sea 21	38) Jeff Kent, SF 22
39) Ray Durham, CWS 20	39) K. Griffey Jr., Cin 22
40) Marty Cordova, Cle 20	40) Dmitri Young, Cin 21

Source: www.espn.com

1. To compare the two leagues construct
    a. Stem and Leaf Plot
    b. Histogram

2. What type of information can you gather from each type of plot?

3. Based on the stem and leaf plots, what can you say about the performance of the two leagues?

4. Based on the histogram, what can you say about the performance of the two leagues?

5. Do the two displays provide the same information? Explain your reasoning.

6. Which display is more informative? Why?

7. From which distribution is it easier to determine the mean number of home runs?

Number Patterns Test

Name: ______________________

Task 1

For each sequence, find the next 3 terms and describe the pattern.

1. 3, 8, 13, 18, 23, ______, ______, ______,...

____

2. 1, 2, 3, 5, 8, 13, ______, ______, ______,...

____

3. 1, 4, 9, 16, ______, ______, ______,...

____

4. 1, 100, 2, 99, 3, 98, ______, ______, ______,...

____

5. 3, -9, 27, -81, ______, ______, ______,...

____

Task 2

For each sequence, find the 10th term and describe the pattern.

1.

Term Number	*1*	*2*	*3*	*4*	...	*10*	
Sequence	*5*	*15*	*25*	*35*	...		

____

2.

Term Number	*1*	*2*	*3*	*4*	...	*10*	
Sequence	*2*	*-4*	*8*	*-16*	...		

____

3.

Term Number	1	2	3	4	5	...	10
Sequence	1	5	11	55	111	...	

___

Task 3

For each sequence, find an expression for the Nth term and describe the pattern.

1.

Term Number	1	2	3	4	...	N	
Sequence	2	7	12	17	...		

___

2.

Term Number	1	2	3	4	...	N	
Sequence	5	25	125	625	...		

___

3.

Term Number	1	2	3	4	...	N	
Sequence	10	18	26	34	...		

___

Toothpick Test

Name: ____________________________

The following toothpick diagrams show a single square constructed from 4 toothpicks and two squares constructed from 7 toothpicks.

One Square

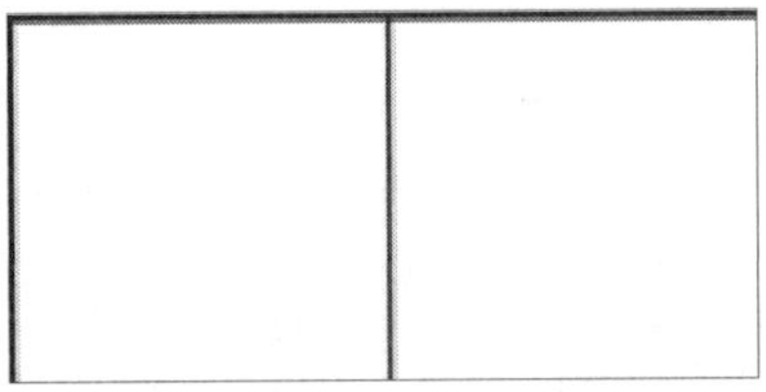

Two Squares

1. Draw toothpick diagrams that show how to construct three squares, four squares, and five squares in the same manner.

   a. Three squares:

   b. Four squares:

   c. Five squares:

2. Complete this table.

Labels	Number of Squares	Number of Toothpicks
Units		
	1	
	2	
	3	
	4	
	5	
	6	
	10	
	50	

3. Use the information in the table and your diagrams to describe the pattern produced by constructing the toothpick squares. Write your answer in complete sentences.

4. Find an expression that gives the number of toothpicks required to construct N squares.

Earning Money

Name: ______________________________

A friend of yours is employed at a company where he makes $5.25 an hour.

Make a table that shows his earnings for various numbers of hours worked (between 1 hour and 20 hours). Include the algebraic expression in your table. Use this information to construct a line graph.

Labels		
Units		
Expression		
	1	
	20	

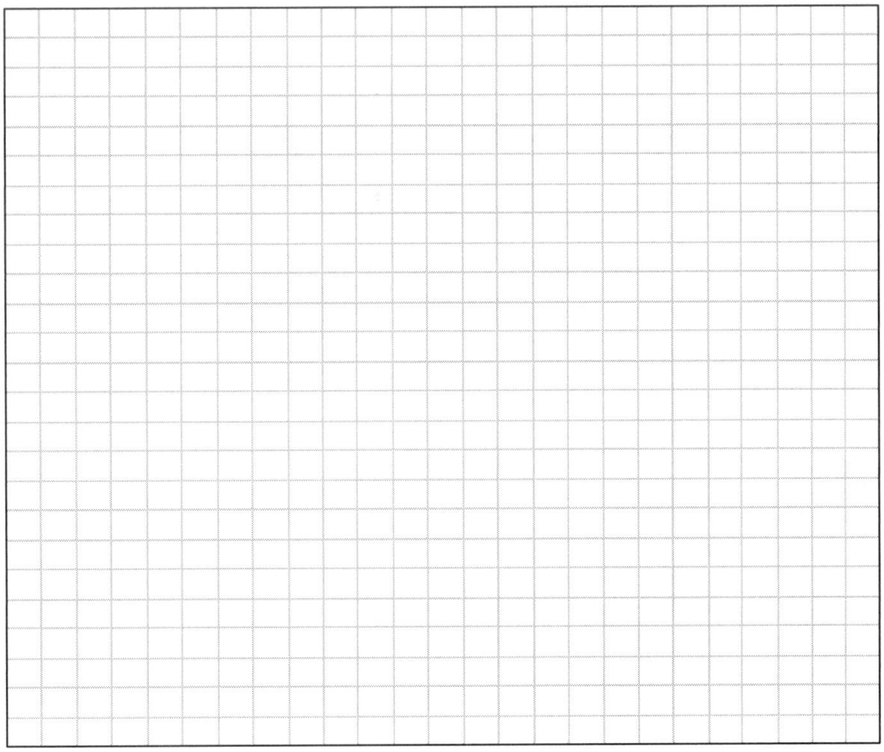

Vacation Flights

Name: ______________________________

You are going on vacation by airplane. The average speed of the plane is 325 miles per hour.

Make a table that shows how far you will travel for various times between 1 hour and 20 hours. Include the algebraic expression in your table. Use this information to construct a line graph.

Labels	***Time in the Air***	***Distance Traveled***
Units		
Expression		
	1	
	20	

Students Who Don't Take Algebra

Name: ______________________________

One out of every four high school students never takes algebra.

Make a table that shows how many students don't take algebra at schools of various sizes. Include the algebraic expression in your table. Use this information to construct a line graph.

Labels		
Units		
Expression		

Paying Taxes

Name: ____________________________

The average family in the United States pays about 38% of their income in taxes of all kinds.

Make a table that shows the amount of taxes paid by families with various incomes. Include the algebraic expression in your table. Use this information to construct a line graph.

Labels		
Units		
Expression		

Stocking the Shelves

Name: ______________________________

You are a volunteer for the school store. One of the most popular items is cool ranch flavored tortilla chips. There are two local vendors that will delivery the chips to the school store. Each makes their deliveries on the first of every month. Chips-R-Us charges 39 cents per bag with a monthly delivery fee of $25. It's In The Bag charges 18 cents per bag with a monthly delivery fee of $65.

1. Write an algebraic equation to model the monthly cost of purchasing chips from Chips-R-Us.

2. Write an algebraic equation to model the monthly cost of purchasing chips from It's In The Bag.

3. Complete the following table summarizing the cost of purchasing chips from each vendor based upon the actual sales by month for the last school year.

Label	**Month**	**Number of bags of chips purchased**	**Cost of purchasing from Chips-R-Us**	**Cost of purchasing from It's In The Bag**
Units				
	September			
	October			
	November			
	December			
	January			
	February			
	March			
	April			
	May			
	June			
Expression				

4. Generate a graph displaying the cost of purchasing from both Chips-R-Us and It's In The Bag.

5. How many bags of chips can you purchase with each vendor if you have the following budgets? Explain how you determined each and write your answer in a complete sentence.

   a. \$50

   b. \$125

   c. \$200

6. Compare the price to purchase from each of the two vendors. Prepare a written analysis indicating which vendor will cost the least and when. Be sure to include when the costs will be the same.

7. If you had to choose one vendor for the entire school year, which would you choose. Would your choice change if you could switch vendors each month? Explain your choices using complete sentences.

Integers and Solving Equations: Form A

Name: ____________________________

Find the answer to each problem without using your calculator.

1. $7 + (-14) =$
2. $-19 + 110 =$
3. $-2 \bullet 10 =$
4. $\frac{210}{-10} =$
5. $13 - 25 =$
6. $-10(-6) =$
7. $-23 + (-14) =$
8. $\frac{-41 + 33}{-2} =$
9. $\frac{10 + (-4) - 4}{2} =$
10. $20(-3) - 10 =$

Solve each equation. Show all your work.

11. $e + 5 = 8$
12. $f + 9 = 13$
13. $-2g = -20$
14. $3h = 18$
15. $20x = -10$
16. $5j = -25$
17. $7 = \frac{k}{-12}$
18. $-2 - L = 18$

Integers and Solving Equations: Form B

Name: ______________________________

Find the answer to each problem without using your calculator.

1. $17 + (-14) =$

2. $-190 + 110 =$

3. $-21 \bullet 10 =$

4. $\frac{-50}{-10} =$

5. $-13 - \cdot 5 =$

6. $-10(-16) =$

7. $-213 + (-124) =$

8. $\frac{-1 - 3}{-2} =$

9. $\frac{10 - 4 - 4}{-2} =$

10. $2(-3) - 10 =$

Solve each equation. Show all your work.

11. $2E + 6 = 10$

12. $4x + 8 = 32$

13. $-2G + 26 = 50$

14. $3h - 18 = 18$

15. $22x + 11 = -11$

16. $3J + (-6) = -15$

17. $12 = 12 + \frac{k}{-7}$

18. $-4 - 4L = 16$

At the Driving Range

Name: ____________________________

Inspired by the saying "practice makes perfect", you decide to perfect your golf swing by going to the local driving range. Unfortunately, you do not have your own set of clubs yet so you must rent from the range. Each bucket of 100 golf balls costs $7.50 and it costs $3.00 to rent clubs.

1. Write an algebraic equation for the total cost at the driving range dependent on the number of buckets of balls you buy. Write your equation using functional notation with C(x) representing the cost.

2. Show all work and answer in complete sentence.
   a. What is the total cost to hit five buckets worth of balls?

   b. How many buckets can you buy with $20?

   c. Find x if C(x) = 28.

   d. Find C(10).

3. Generate a graph displaying the total cost of practicing at the driving range

4. The owner of the driving range is looking at the receipts for the day. Are the following costs correct? Explain why or why not in a complete sentence.

   a. $48 for 6 buckets and club rental

   b. $70 for 9 buckets and club rental

5. What would be the domain based on the problem situation?

6. Based on your answer to question 5, what would be the range according to the problem situation?

7. Set up an inequality to solve each. Graph the solution to each.

   a. Determine the number of buckets of balls you could hit if you had to spend at most $50 but at least $25.

   b. $C(x) > 55$ or $C(x) <= 40$

Balloon Problem

Name: ____________________________

Here you will examine a problem situation about buying and selling helium balloons. Show all your work and write your answers in complete sentences.

You have the opportunity to run the helium balloon concession at a local fair. In order to sell balloons, you must pay a $25 fee to the fair sponsors and buy $75 worth of balloons and helium. You can sell each balloon for $0.80. You need to calculate your net profit for the fair. Remember, profit is the money you bring in minus the expenses you have to pay out.

1. How much profit will you have if you sell:

   a. 100 balloons?

   b. 200 balloons?

   c. 500 balloons?

   d. 50 balloons?

   e. 80 balloons?

   Write a complete sentence describing how you found these answers.

2. Do your answers in 1d and 1e *make sense*? Why or why not?

3. Algebraic Expression or Equation for Profit ________________________________

4. How many balloons must you sell to make a *profit* of:

   a. $40?

   b. $100?

   Write a complete sentence describing how you found these answers.

5. How many balloons must you sell to post the following *losses?* (Note: A loss is a negative number. For example, a loss of $20 is -$20.)

   a. $20

   b. $50

   c. $100

   Write a complete sentence describing how you found these answers.

6. Use the information you have found so far to complete this table.

	Balloons Sold	Profit
Labels	***Balloons Sold***	***Profit***
Units		
Expression		
	200	
	500	
	50	
		80
		40
		100
		-20
	10	
	0	
	300	
	400	
		120
		150
		-4

7. Use the information from the table to construct a graph that illustrates this problem situation.

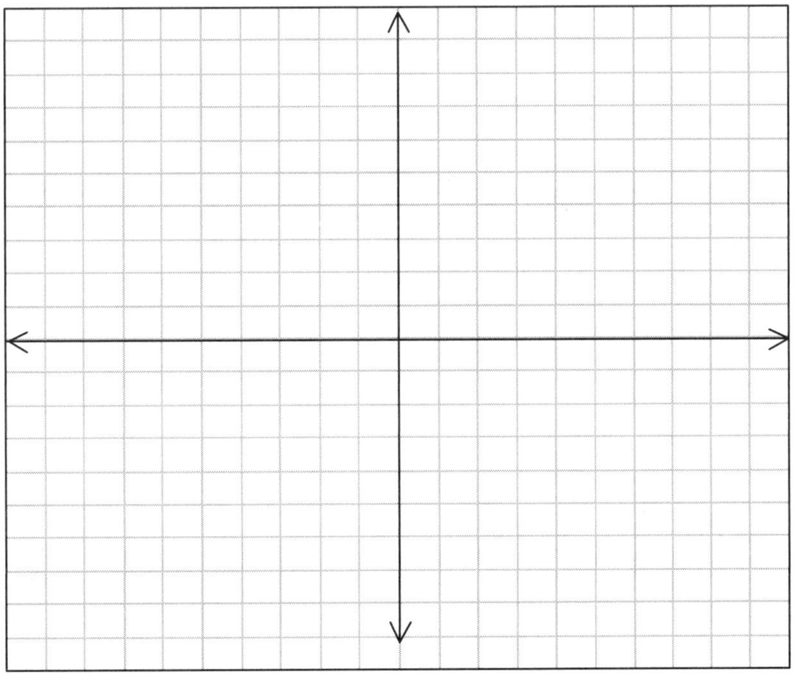

8. How many balloons do you have to sell in order to *break even*? Justify your answer, and write a complete sentence describing how you found this answer.

Television Problem and Climber Problem

Name: ___________________________

Television Problem

Use this graph to answer the questions about televisions.

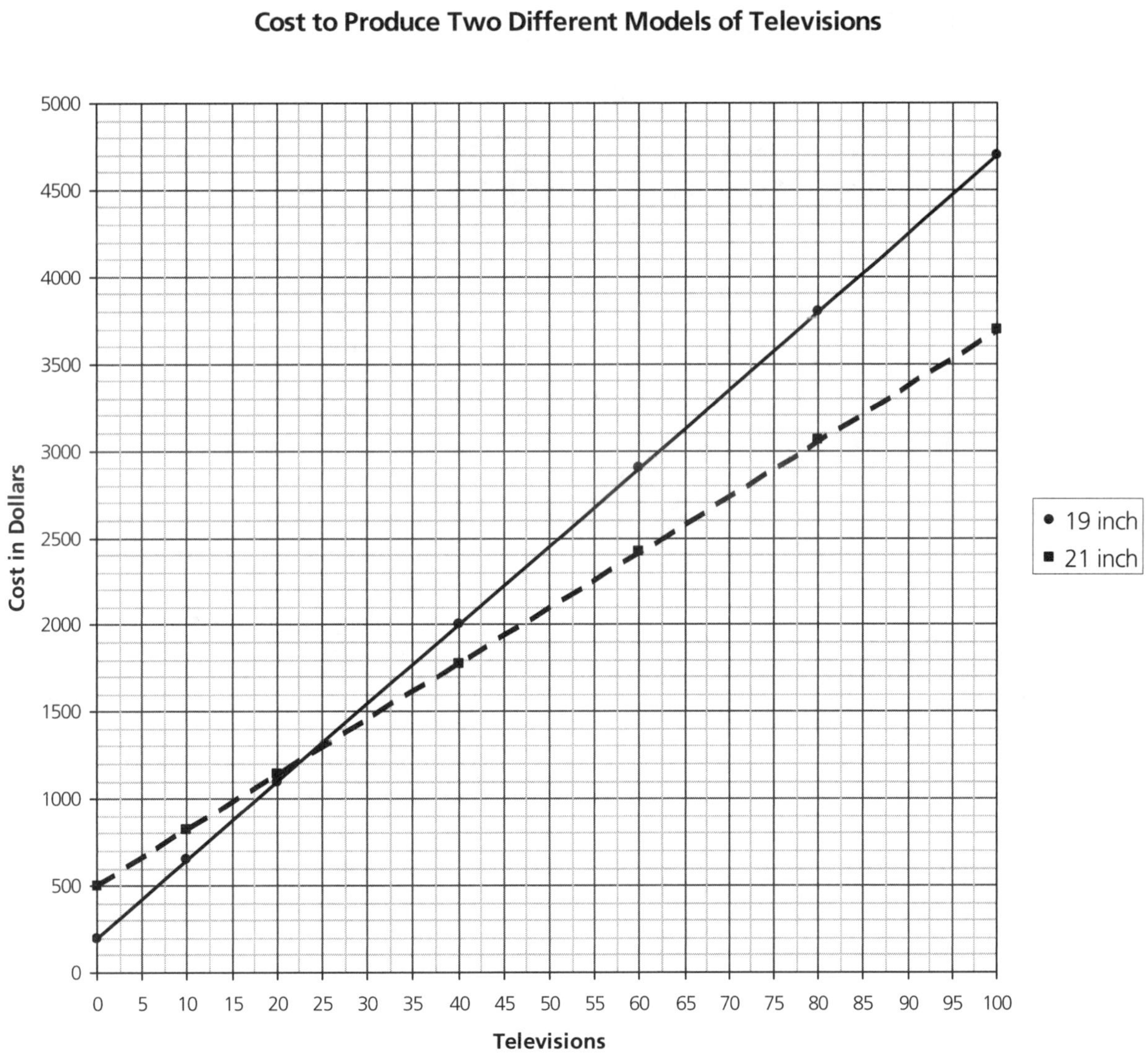

1. What information does this graph show?

2. How much does it cost to produce:
   a. thirty-five 19-inch televisions?

   b. twenty-five 19-inch televisions?

   c. thirteen 19-inch televisions?

3. How much does it cost to produce:
   a. forty-five 21-inch televisions?

   b. fifteen 21-inch televisions?

   c. eight 21-inch televisions?

4. How many of each type of television must be produced for the costs to be the same? How do you know?

5. Use the graph to predict the cost of producing:

   a. sixty 19-inch televisions.

   b. sixty 21-inch televisions.

   c. one hundred 21-inch televisions.

   d. one hundred 19-inch televisions.

Extra credit:

Find an algebraic expression or equation for the cost to produce 19-inch televisions. Find another algebraic expression or equation for the cost to produce 21-inch televisions.

Rule for producing 19-inch televisions: ______________________________

Rule for producing 21-inch televisions: ______________________________

Climber Problem

A mountain climber is currently 77 feet above ground level on the side of a cliff. She is climbing at the rate of about 7 feet per minute.

Answer the following questions in complete sentences.

1. How many feet above ground level will she be in:
   a. 10 minutes?

   b. 25 minutes?

2. Currently, how many feet above ground level is she?

3. How many feet above ground level was she:
   a. 5 minutes *ago*?

   b. 10 minutes *ago*?

   c. 15 minutes *ago*?

4. How many minutes *ago* did she start her climb?

5. When will she be:
   a. 154 feet above ground level?

   b. 175 feet above ground level?

   c. 287 feet above ground level?

6. The top of the cliff is 450 feet above ground level. When will she reach the top of the cliff?

7. Write an algebraic rule for the climber's height above ground level.

8. Fill in this table and construct a graph for this problem.

Labels		
Units		
Expression		

Widgets-R-Us

Name: ______________________________

Widgets cost $4 each, plus $10 per order for shipping and handling.

For this problem situation, find an appropriate algebraic equation, define each variable, make a table of values, construct a graph, and use this information to answer the questions.

1. Find an algebraic equation for this situation.

   Equation: ______________________________

   Define each variable in your equation by writing a short phrase that describes what the variable represents.

   x = ______________________________

   y = ______________________________

2. Make a table with at least 5 values of x and y.

x	y
-2	
-5	
0	

3. Graph the equation for this problem situation.

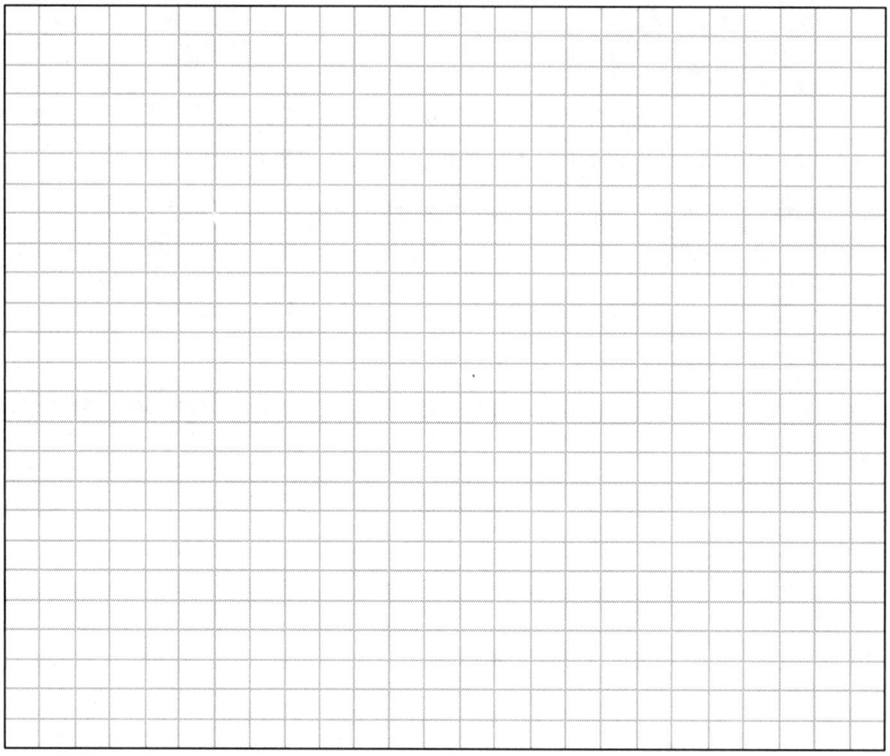

Use your graph, your equation, or your table to answer the following questions. Write your answers in complete sentences.

4. How would you describe your graph?

5. Where does the graph intersect the *x*-axis? Why?

6. Where does the graph intersect the *y*-axis? Why?

7. What is the cost if you order:
   a. 7 widgets?

   b. -9 widgets?

8. How many widgets did you order if the total cost is:
   a. $66?

   b. -$10?

9. Write an equation that states: the cost of an order of widgets is $212. Solve this equation to find how many widgets you ordered.

10. Find the cost if you order -23,456 widgets.

11. Do problems 9 and 10 *make sense* in this problem situation? Why or why not?

Computer Rental Problem

Name: ____________________________

For a college computer class in which you are enrolled, you must rent a computer. Two rental companies carry the particular model you need.

Computronics rents the computer you need for $2.50 a day plus a one-time registration fee of $27. A second company, Computers for Students, rents the same computer at a rate of $4.75 per day.

You know you will need the computer for 30 days or less, but you don't know exactly how many days. From whom should you rent? In order to answer this question, write the problem situation for each company, construct a table of values, and find the algebraic expression. Then draw graphs for both companies on a single grid.

Task 1

Write the problem situation for Computronics. Construct a table of values and find the expression.

Problem Situation:

Labels		
Units		
Expression		

Task 2

Write the problem situation for Computers for Students. Construct a table of values and find the expression.

Problem Situation:

Labels		
Units		
Expression		

Task 3

Draw graphs that show the information from each company.

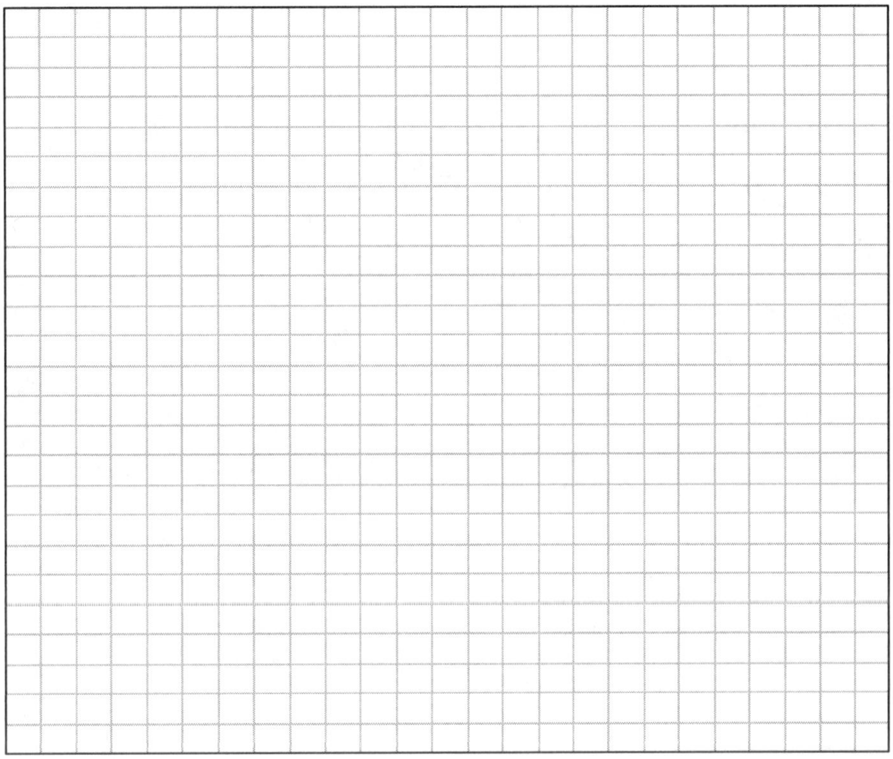

Task 4

Use the information you have gathered and analyzed in Tasks 1– 3 to write a paragraph explaining which company you will rent from and why. Write your answer in complete sentences.

Rectangle Area Quiz

Name: ___________________________

1. These rectangular boxes need to be wrapped. To find out how much wrapping paper you need, calculate the total surface area of each box. (Hint: Each box has six sides.)

a.

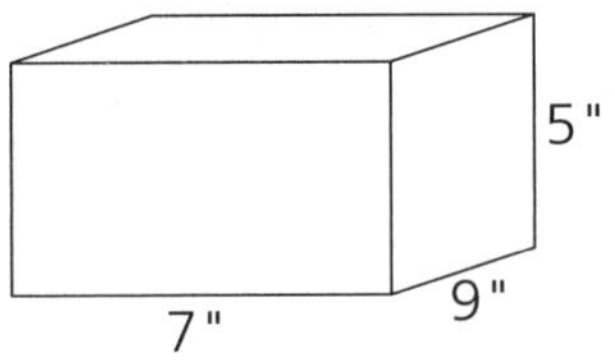

b.

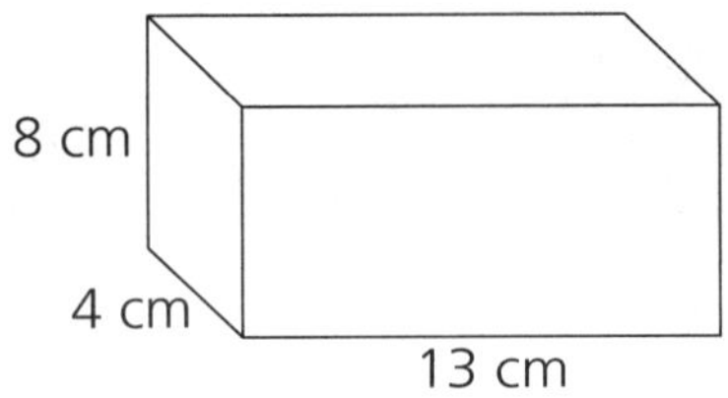

c.

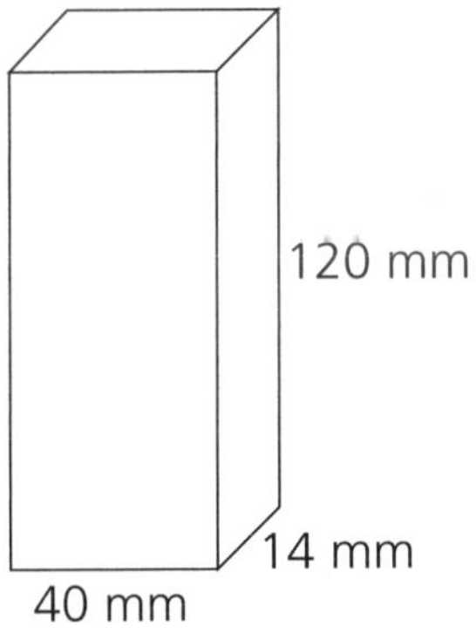

2. In the figure, a side of the inner square measures 20 inches. A side of the outer square measures 30 inches. Find the area of the shaded region.

3. In the figure, the outer rectangle measures 8 feet by 12 feet. The inner rectangle measures 5 feet by 7 feet. Find the area of the shaded region.

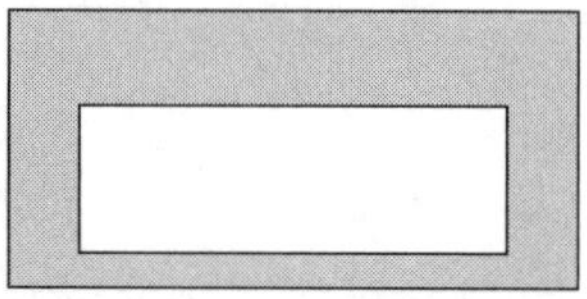

4. Measure the figure with a centimeter ruler to find the area of the shaded region.

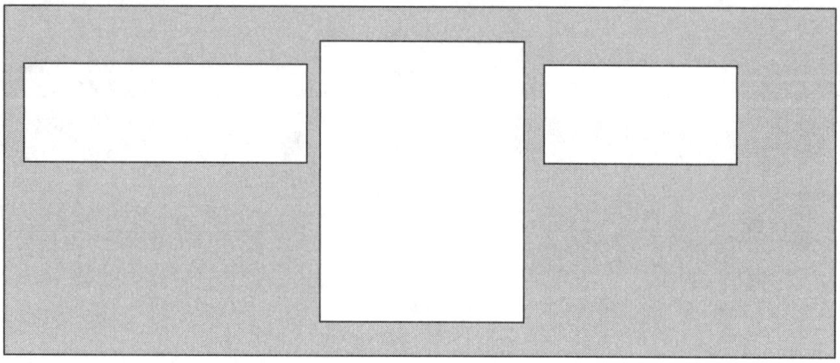

5. The area of a rectangular field is 680 square miles. If the length of the field is 80 miles, find the width of the field. Write a sentence explaining how you arrived at your answer.

6. The area of a square garden is 529 square feet. Find the perimeter of the garden. Write a sentence explaining how you arrived at your answer.

7. You just bought a 50-square-foot roll of wrapping paper. The width of the roll is 1 foot. What are the dimensions of the largest rectangular box you can wrap with this roll of paper? Write a sentence explaining how you arrived at your answer.

8. You are designing a red and black checkerboard banner, which will have an area of 576 square inches. All checkerboards have eight rows and eight columns of squares. How long is the rod you will need to hang the banner? How many square inches of red cloth will you need to make the banner? Write a sentence explaining how you arrived at your answer.

Parallelogram Area Quiz

Name: ____________________________

1. Find the area of each parallelogram.

   a. 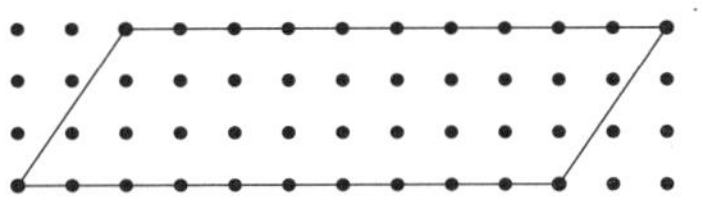

   b. 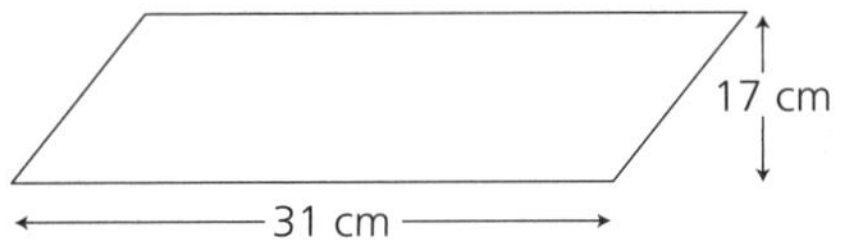

   c. 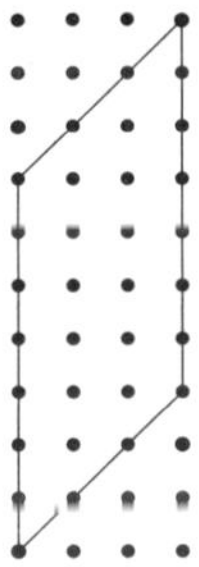

   d. 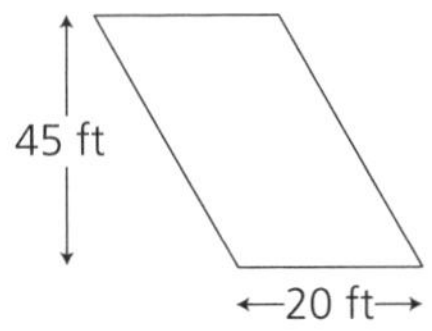

   e. 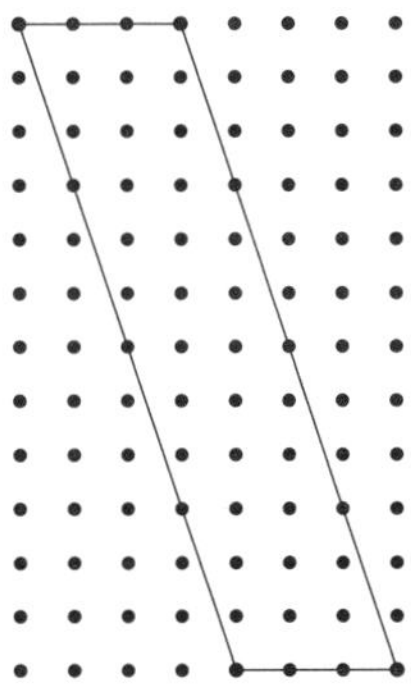

2. If the height of a parallelogram is 22 inches and the length of the base is 25 inches, find the area of the parallelogram. Show your work.

3. If the base and the height of a parallelogram are equal in measure, the parallelogram must be a square. Is this statement true or false? Explain your answer.

4. If the base and height of a parallelogram are doubled in size, the area will also double in size. Is this statement true or false? Explain your answer.

5. The area of a parallelogram is 330 square feet and the base measures 22 feet. Find the height of the parallelogram. Show your work.

6. If the area of a parallelogram is 390 square yards and the height is 15 yards, find the length of the base of the parallelogram. Show your work.

7. The area of a parallelogram is 200 square inches and the base is twice the measure of the height. Find the base and the height of the parallelogram. Show your work.

8. If the area of a parallelogram is 192 square inches and the base is three times the measure of the height, find the base and the height. Show your work.

Triangle Area Quiz

Name: ______________________________

1. Find the area of both triangles.

   a. 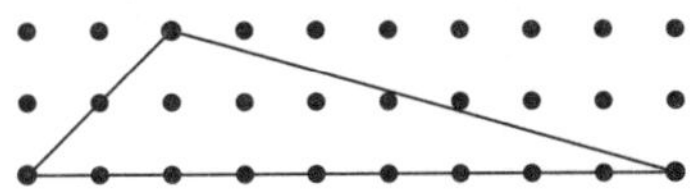

   b. 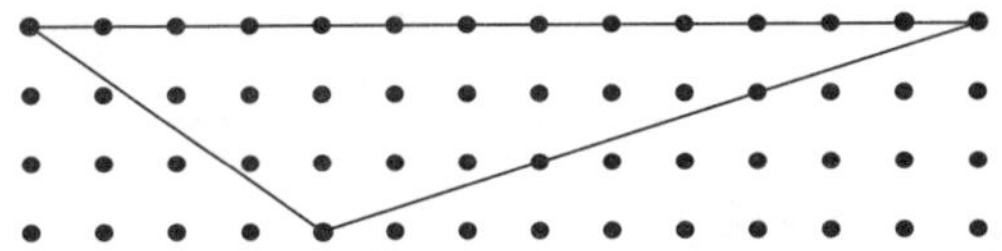

2. Find the area of each figure.

   a. 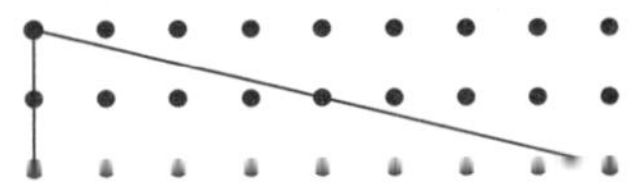

   b. 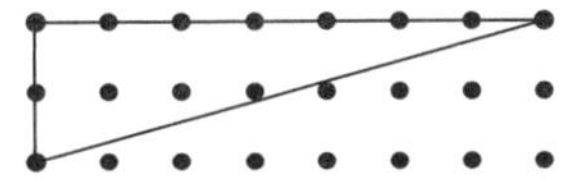

   c. 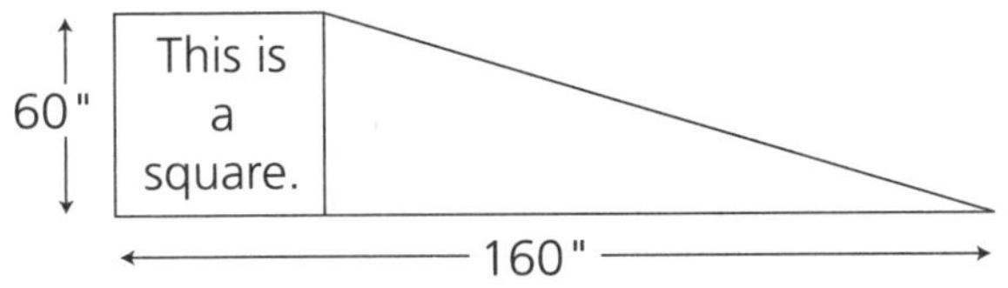

3. You are making a nylon kite. The kite sticks are perpendicular to each other, and the longer stick intersects the middle of the shorter stick. If the height of the kite is 42 inches and its greatest width is 30 inches, find the amount of nylon you will need to make the kite. (Hint: Find the area.)

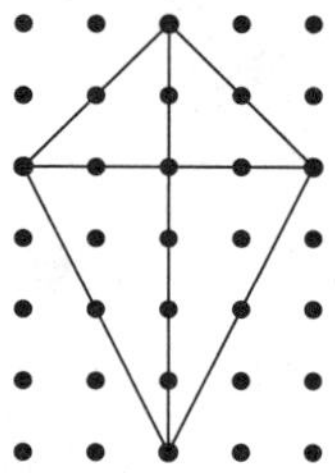

4. You are designing a triangular pennant for your school. If you have only 330 square inches of fabric and you want the pennant, when suspended from its base, to hang down 44 inches, what will be the length of the base?

5. The perimeter of a baseball diamond is 360 feet. (A baseball diamond is actually a square.) How many square feet of grass must be mowed inside the triangle formed by the pitcher's mound, second base, and third base? Write a sentence explaining how you arrived at your answer.

6. The sail on a boat is triangular and required 540 square feet of nylon to make. If the base of the sail is 36 feet, find its height.

7. A triangular window in a beach house is 48 inches in height and contains 720 square inches of glass. Find the measure of the base of the window.

Trapezoid Area Quiz

Name: ____________________________

1. Find the area of each trapezoid.

a.

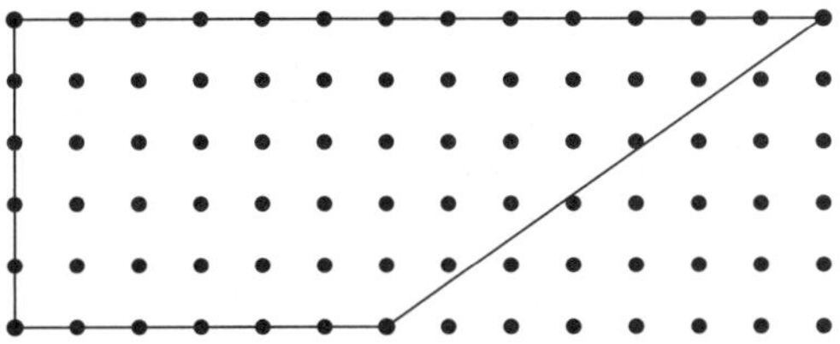

b.

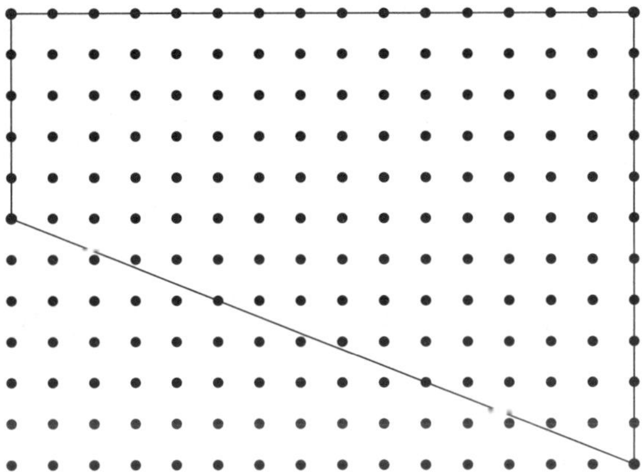

c.

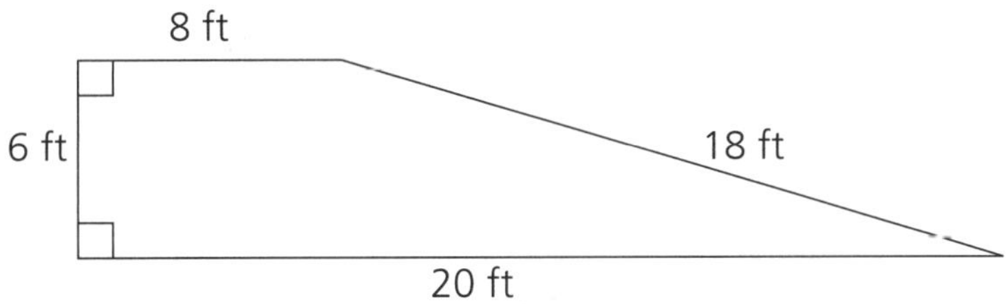

d.

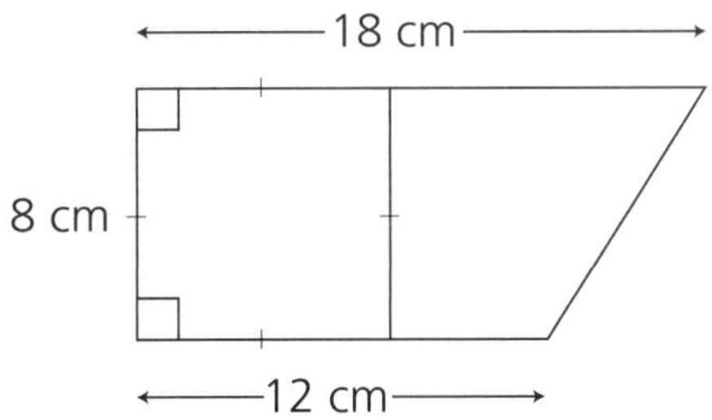

2. If the height of a trapezoid is 5 inches, and the bases measure 6 inches and 11 inches, find the area of the trapezoid. Show your work.

3. Can the bases of a trapezoid ever be the same length? Explain why or why not.

4. If the bases of a trapezoid measure 20 feet and 42 feet, and the height is 24 feet, find the area of the trapezoid.

5. The area of a trapezoid is 396 square yards, and the bases measure 13 yards and 20 yards. Find the height of the trapezoid. Show your work.

6. The area of a trapezoid is 323 square miles, one base measures 26 miles, and the height is 24 miles. Find the length of the second base of the trapezoid. Show your work.

7. The area of a trapezoid is 150 square meters. The height is 20 meters, and one base is three meters longer than the other base. Find the length of both bases. Show your work.

8. The area of a trapezoid is 504 square feet and the height is 12 feet. Find the length of both bases if the first base is twice the length of the second base. Show your work.

Polygon Area Quiz

Name: ______________________

1. Find the area of each regular polygon.

a.

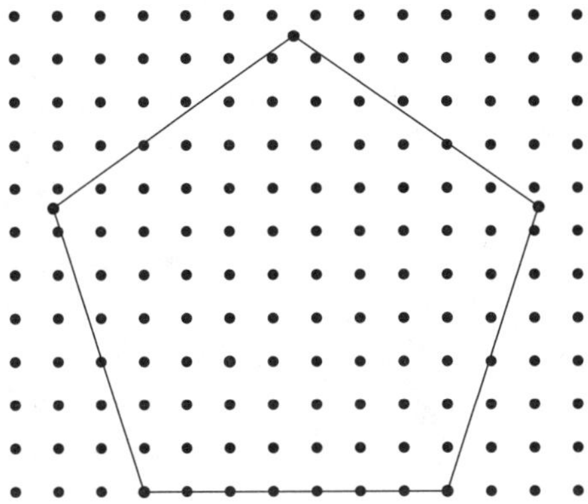

b.

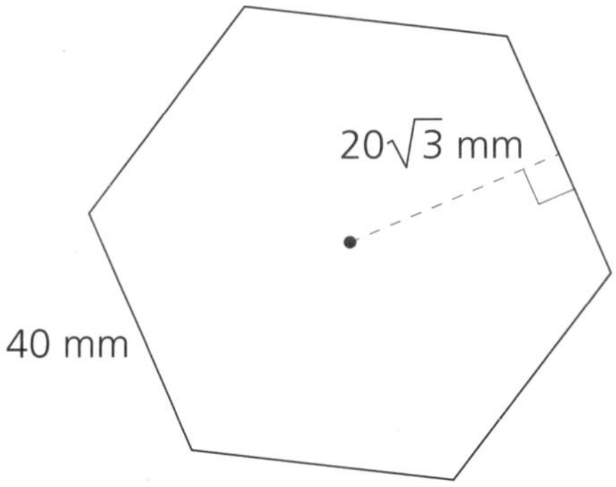

c.

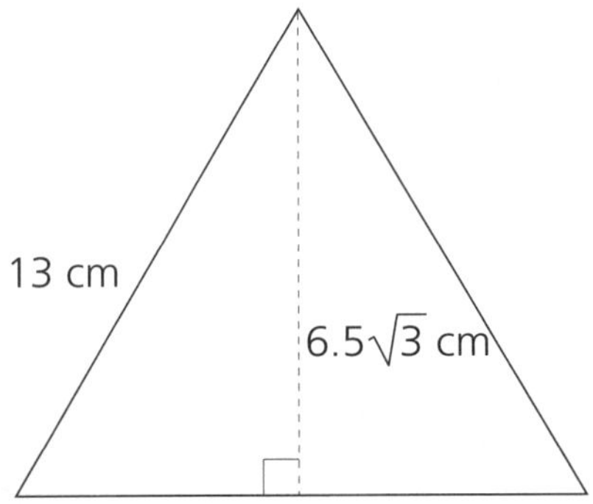

2. If the distance from the center to a side of a regular heptagon is 20.76 inches and each side measures 20 inches, find the area of the heptagon. Show your work.

3. If the distance from the center to a side of a stop sign is 36.21 inches and the length of each side is 30 inches, find the area of the stop sign. Show your work.

4. If the apothem (distance from the center to a side) of a regular decagon is 33.85 meters and each side measures 22 meters, find the area of the decagon. Show your work.

5. If the area of a regular nonagon is 1211.49 square yards and the apothem is 19.23 yards, find the length of each side. Show your work.

6. A quilt is made from interlocking regular hexagons. The distance from the center to a side of each hexagon is 1.7 inches and the perimeter of each hexagon is 12 inches. How many regular hexagons would be needed to make a quilt measuring 8 feet by 8 feet? Show your work.

7. If the side of a square is 20 meters long, what is:

   a. The area of the square?

   b. The distance from the center to a side (apothem)?

   c. If the length of a side doubles how does that affect the area of the square? (Does it double?) How does it affect the length of the apothem?

Circle Area Quiz

Name: ____________________________

1. Find the area of each figure. (What appears to be a semicircle is just that.)

a.

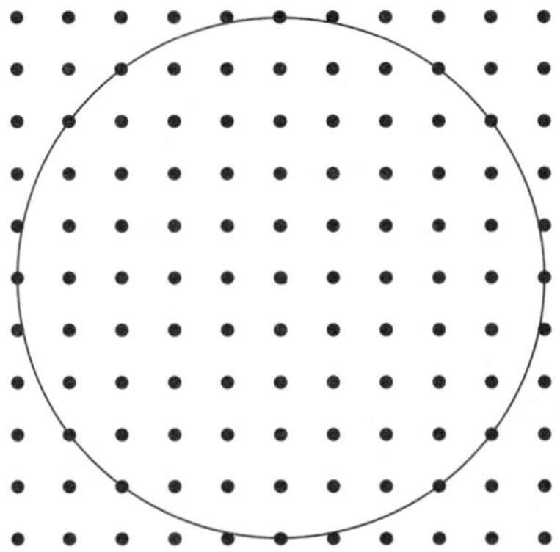

b.

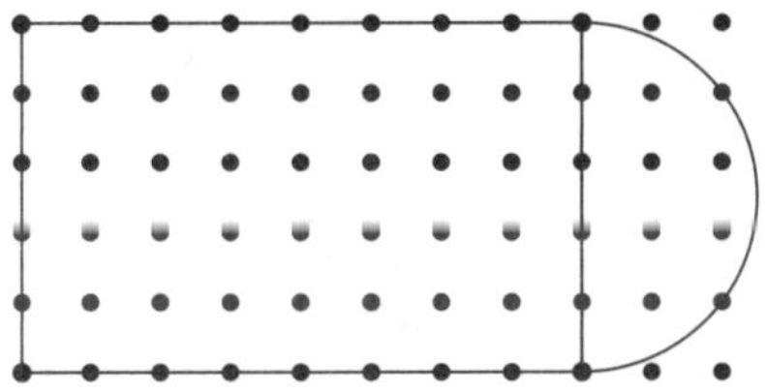

c.

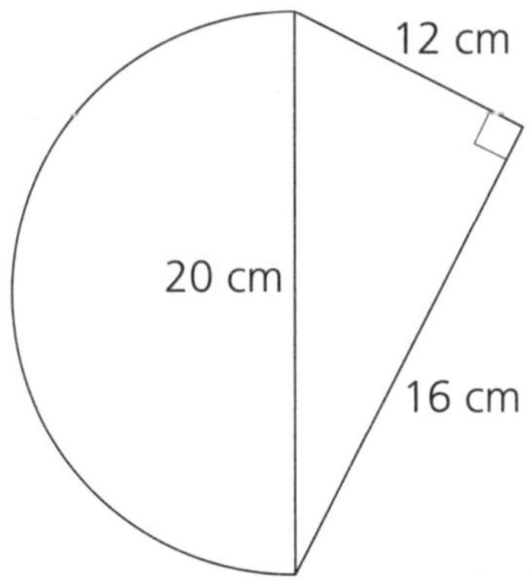

2. A spotlight rotates 360° and emits a light beam that is visible up to 1500 feet from the light. Find the area covered by the light. Show your work.

3. A lawn sprinkler spins 360° and has a 4-foot spray. Find the area of lawn the sprinkler will water.

4. A circular bed of flowers has a circumference of 8 feet. Find the area of the flower bed.

5. In the figure, the area of the square is 36 square feet. The diagonal of the square is $\sqrt{72}$ feet. Find the area of the shaded region.

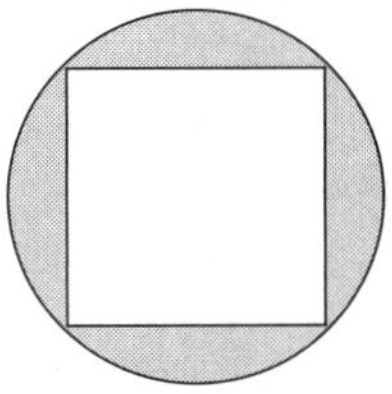

6. If it takes 28.26 square inches of plastic to manufacture the circular lid for a paper cup, what is the radius of the cup? Show your work.

7. In a circular baby pool, 3215.36 square feet of water is exposed to the air. To make a cover for the pool you need to know its diameter. Find the diameter of the pool. Show your work.

8. In this figure, the area of the square is 121 square feet. The circle is just touching each side of the square. Find the area of the shaded region. Show your work.

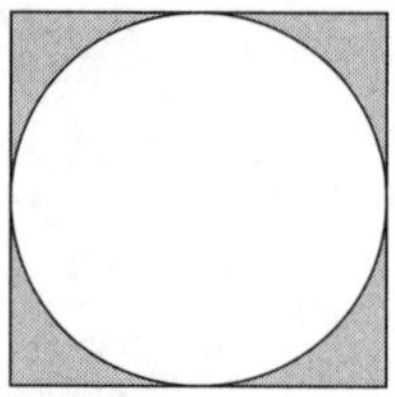

9. This figure is composed of a semicircle and a square. The area of the square is 169 square feet. Find the perimeter of the figure.

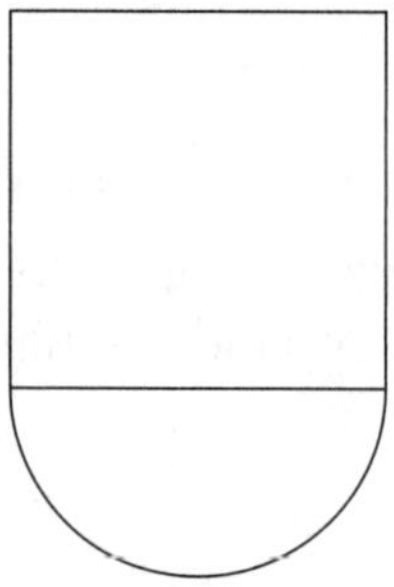

Pythagorean Theorem Test

Name: ___________________________

For problems 1–4:

- Draw a picture
- Show your work
- Write a complete sentence stating your solution

1. If a small rectangular box measures 10 cm by 5 cm by 2 cm, what is the longest toothpick that will fit in the box without bending?

2. A box of long-stemmed roses just arrived at your house. If the rectangular box measures 3 feet by 1 foot by 0.5 feet, find the maximum possible length of one rose. (Assume that the rose and the stem of the rose are not bent.)

3. A square sheet of glass measures 8 feet on each side. Describe what will happen if you try to get it through a rectangular door frame that measures 4 feet by 7 feet. Mathematically prove whether the glass *will* or *will not* fit through the door. (Do not concern yourself with the thickness of the glass.)

4. A letter measuring 33 cm by 4.9 cm was in your mailbox today. If your mailbox measures 30 cm by 15 cm by 5 cm, did the letter fit entirely in the mailbox without folding or bending?

For problems 5–10, each triangle is a right triangle. Use the Pythagorean Theorem to find the length of the unknown side. Show all or your work in solving each problem.

5.

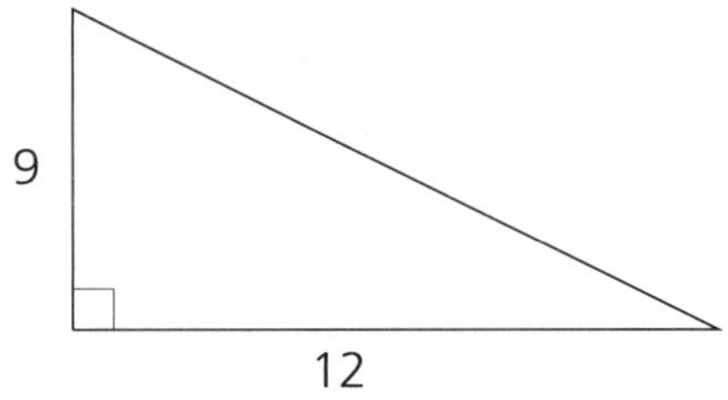

6.

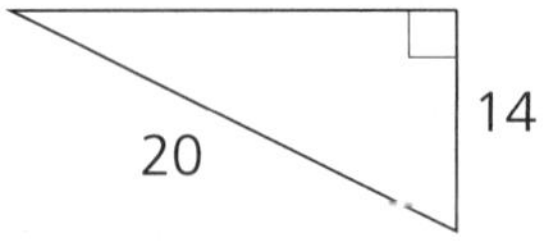

7.

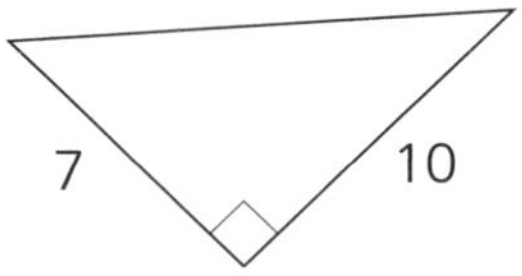

8.

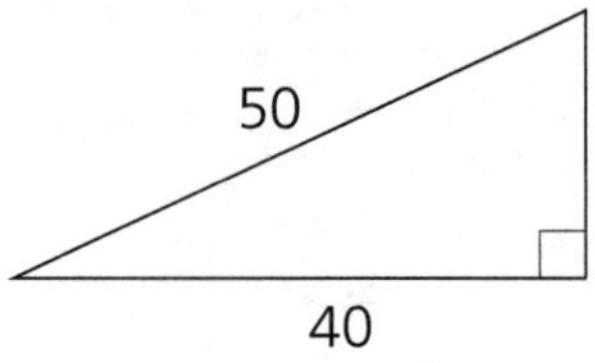

9.

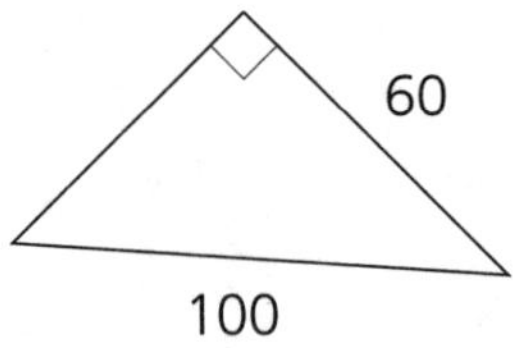

10.

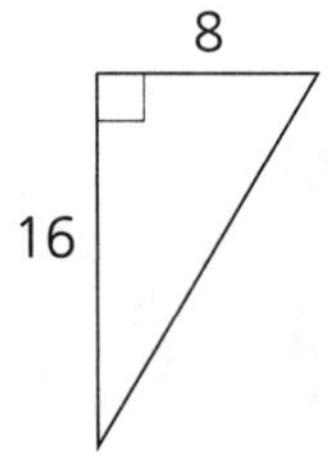

Special Right Triangle Quiz 1 (Form A)

Name: ___________________________

Show the work required to solve each problem.

1. If the legs of this isosceles triangle each measure 24 inches, find the length of the hypotenuse.

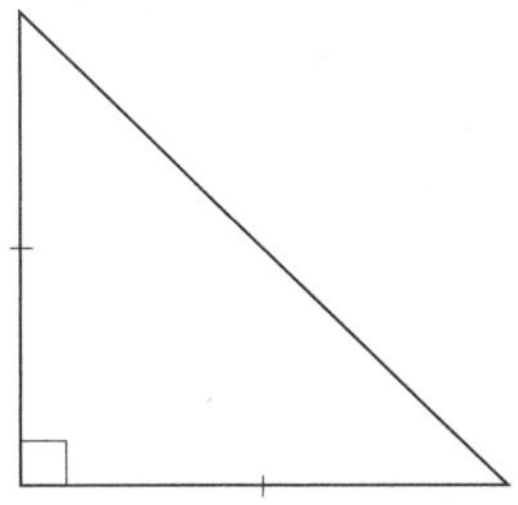

2. Find a.

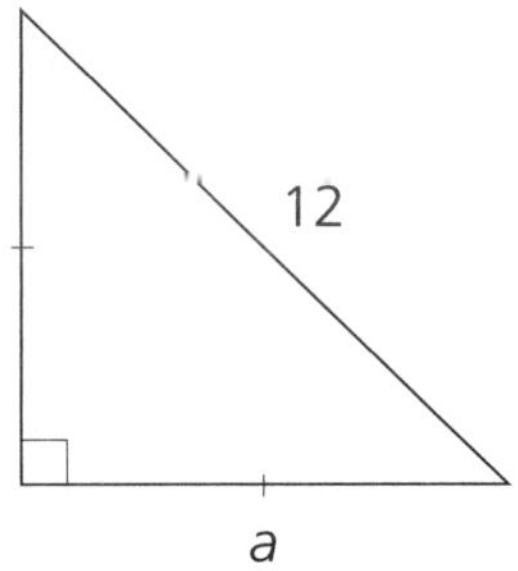

3. The area of this square is 36 cm^2. Find the length of its diagonal.

Special Right Triangle Quiz 1 (Form B)

Name: ______________________________

Show the work required to solve each problem.

1. If the hypotenuse of this isosceles triangle measures 24 inches, find the length of the each leg.

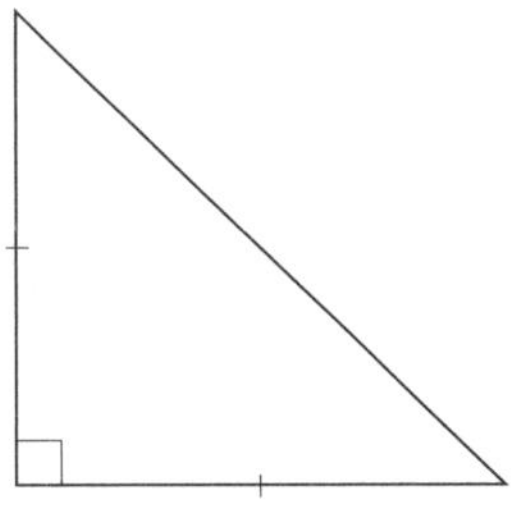

2. Find *a*.

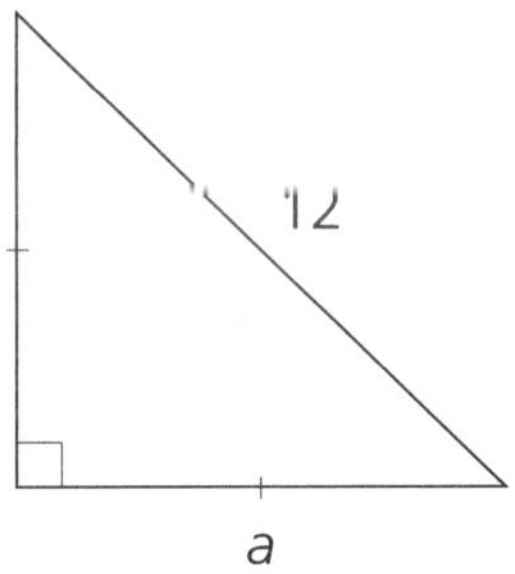

3. If the diagonal of this square measures 20 cm, find its area.

Special Right Triangle Quiz 1 (Form C)

Name: ____________________________

Show the work required to solve each problem.

1. Find the area of this figure using the information given on the diagram. (The figure is composed of a right triangle and a semicircle.)

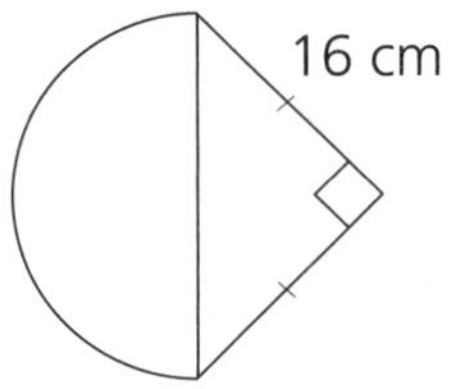

2. If the area of the square is 121 square inches, find the perimeter of this figure. (The figure is composed of a square and a semicircle.)

Special Right Triangle Quiz 2 (Form A)

Name: ______________________

Show the work required to solve each problem.

1. If the measure of the hypotenuse in this 30-60-90 triangle is 18 meters, find the length of sides *a* and *b*.

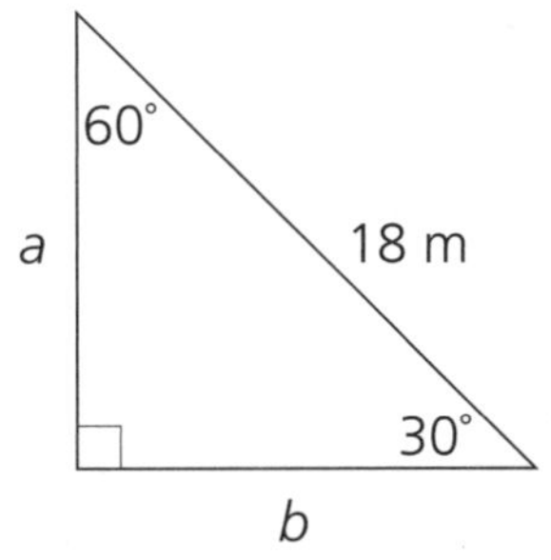

2. In the triangle below, if the measure of the side opposite the 30° angle is 15 feet, find the length of sides *b* and *c*.

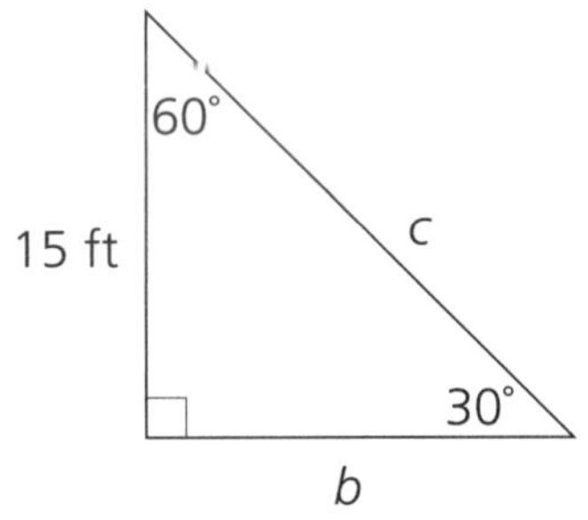

3. In the triangle below, if the measure of the side opposite the 60° angle is 9 millimeters, find the length of sides *a* and *c*.

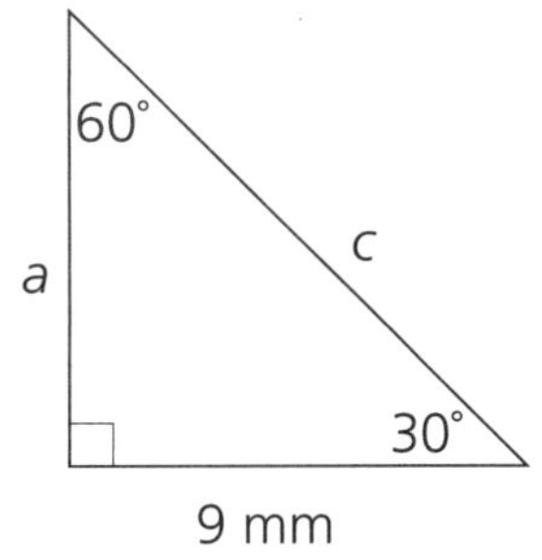

4. A microwave dish is situated on top of a tower, and the signal travels from the dish to your house so you can watch cable TV. The angle of elevation from your house to the tower measures 30° and the distance from your house to the tower is 22 miles. Find the height of the tower and the distance the signal travels.

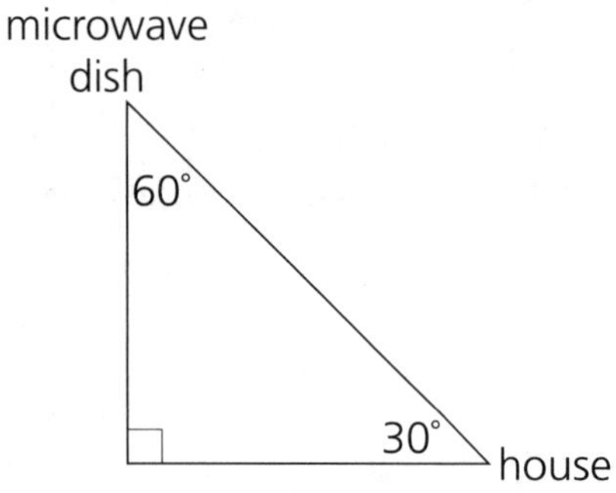

5. The measure of the longest leg in this 30-60-90 triangle is 42 miles. Find the length of the hypotenuse.

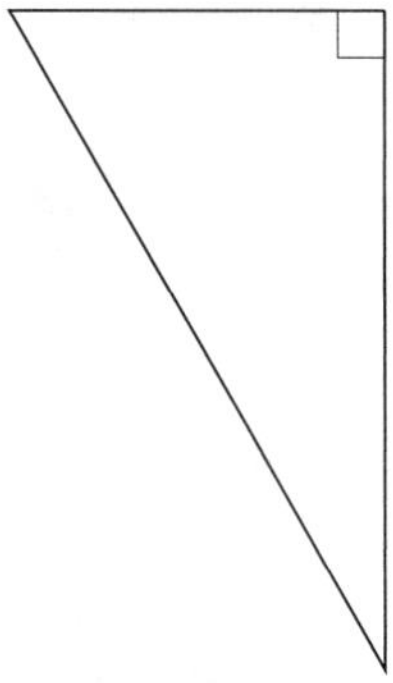

6. The length of the shortest leg in this 30-60-90 triangle is 33 meters. Find the length of the hypotenuse.

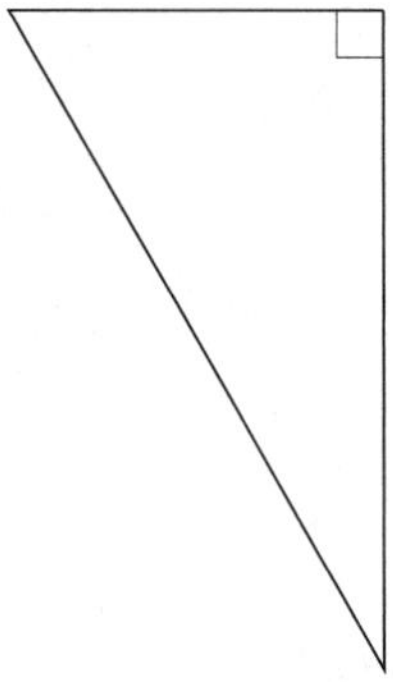

Special Right Triangle Quiz 2 (Form B)

Name: ____________________________

Show the work required to solve each problem.

1. If the measure of the longest leg in this 30-60-90 triangle is 10 meters, find the length of sides *a* and *c*.

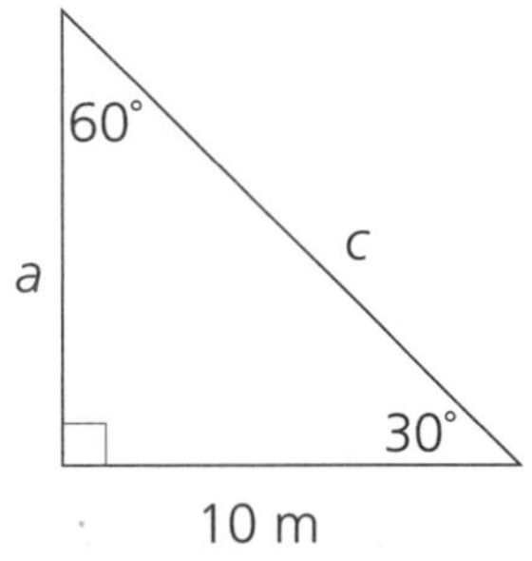

2. If the measure of the shortest leg of this 30-60-90 triangle is 25 feet, find the length of sides *b* and *c*.

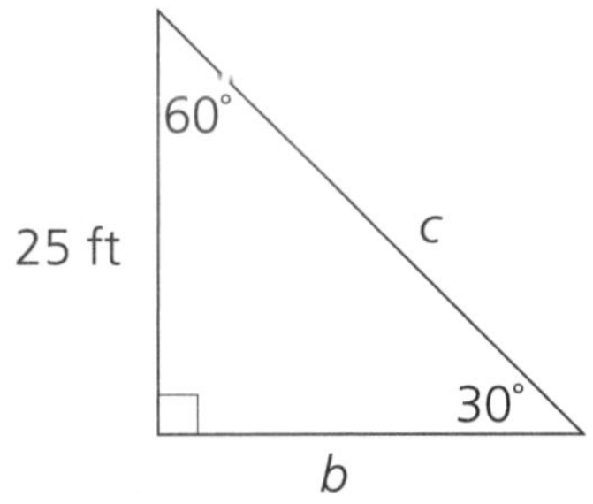

3. If the measure of the hypotenuse in this 30-60-90 triangle is 9 mm, find the length of sides *a* and *b*.

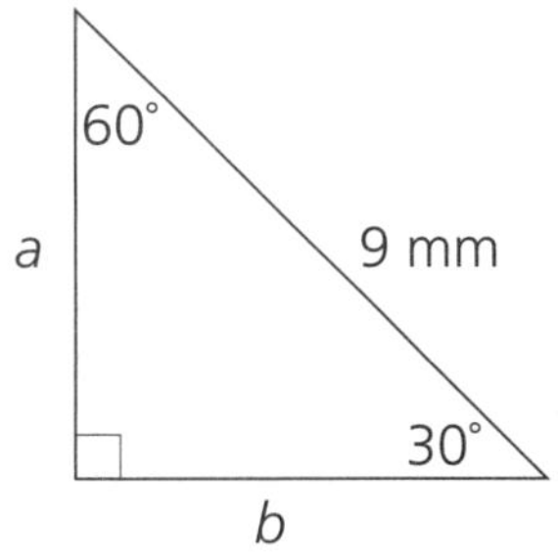

4. A microwave dish is situated on top of a 50-foot tower; the signal travels from the dish to your house so you can watch cable TV. The angle of elevation from your house to the dish measures 30°. Find the distance from your house to the base of the tower and the distance the signal travels.

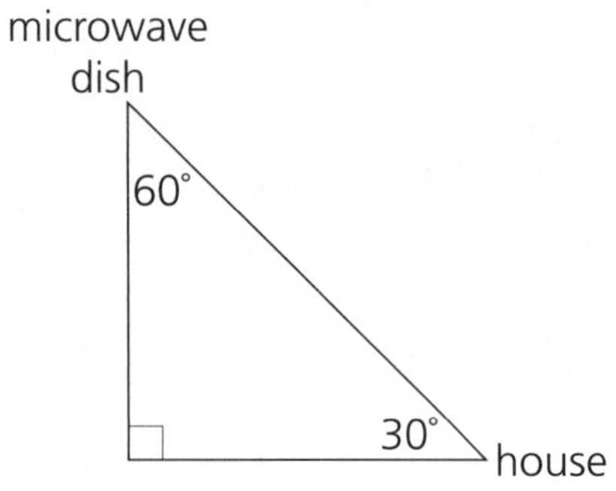

5. The measure of the shortest leg in this 30-60-90 triangle is 42 meters. Find the length of the hypotenuse.

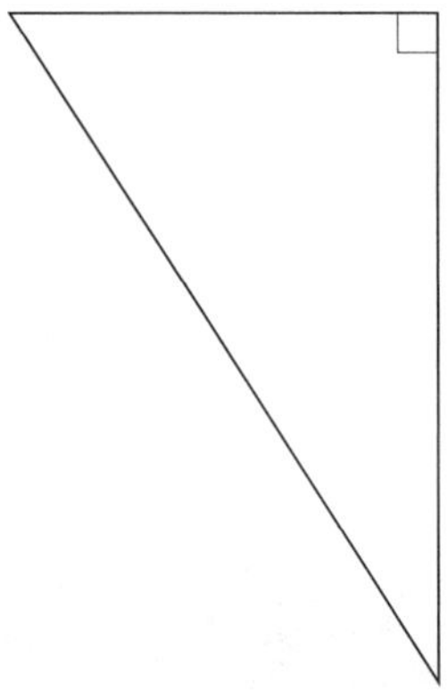

6. The measure of the longest leg in this 30-60-90 triangle is 33 meters. Find the length of the hypotenuse.

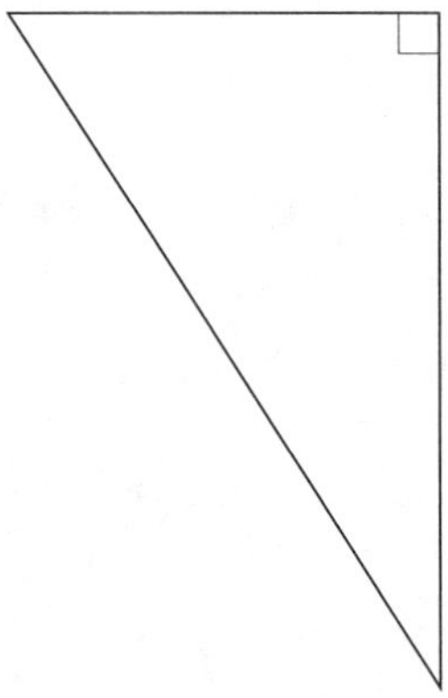

Special Right Triangle Quiz 2 (Form C)

Name: ____________________________

Show the work required to solve each problem.

1. If the area of this trapezoid is 156 cm^2, find its perimeter.

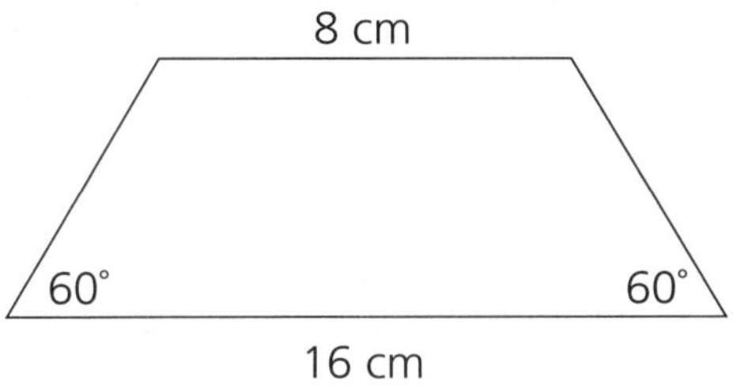

2. Find the area of this triangle.

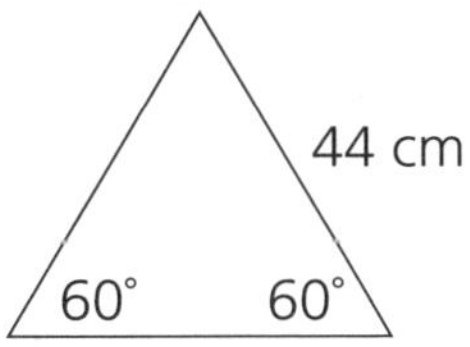

3. Find the area of this trapezoid.

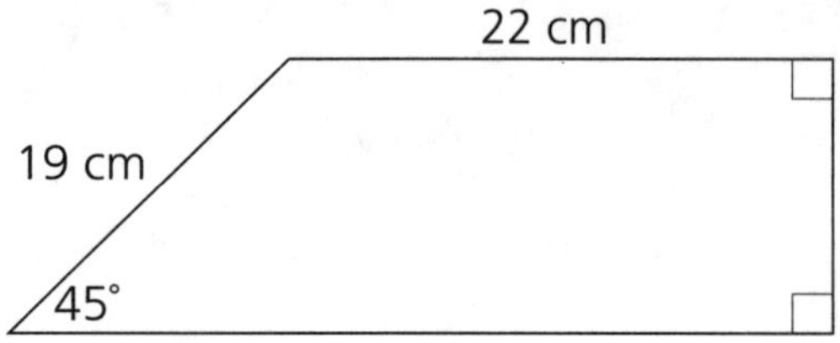

Special Right Triangle Quiz 2 (Form D)

Name: ____________________________

Show the work required to solve each problem.

1. If the area of this trapezoid is 130 cm^2, find its perimeter.

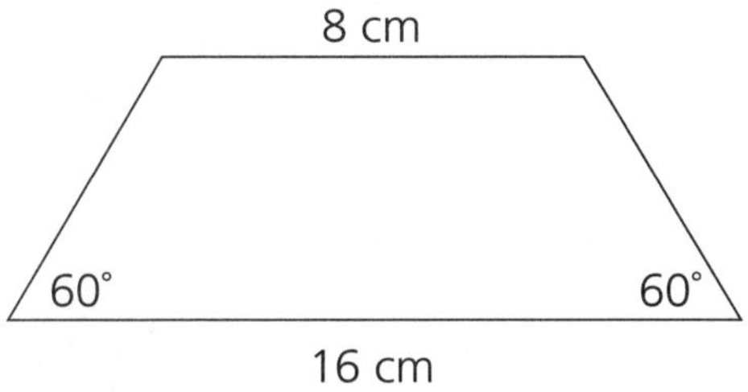

2. Find the area of this triangle.

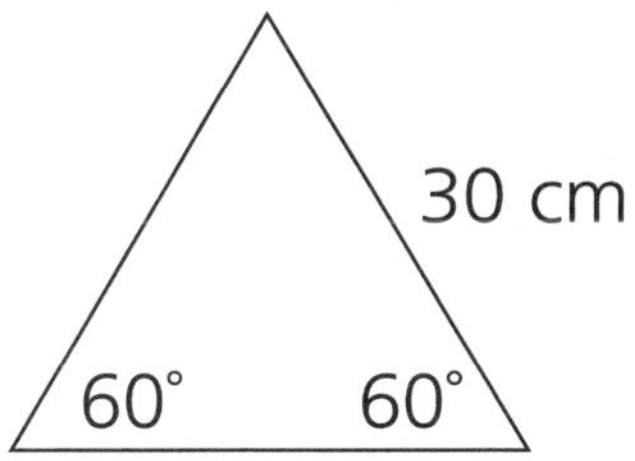

3. Find the area of this trapezoid.

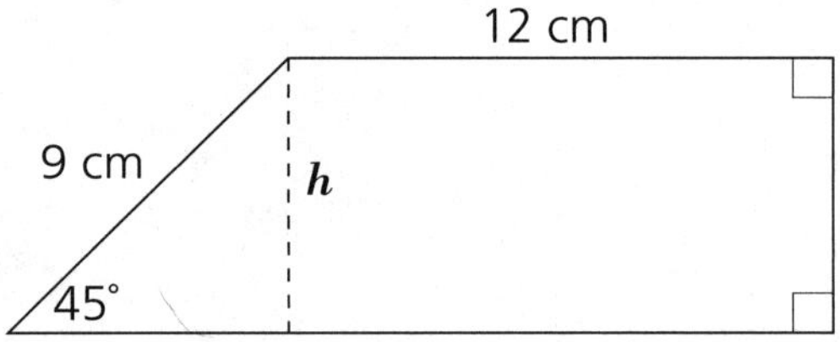

Quiz: Right Triangle Trigonometry 1

Name: ______________________________

The following statement is the ADA Title III requirement for wheelchair ramps.

The least possible slope shall be used for any ramp. The maximum slope of a ramp in new construction shall be 1:12. The maximum rise for any run shall be 30 inches. —Title III, Section 4.8.2

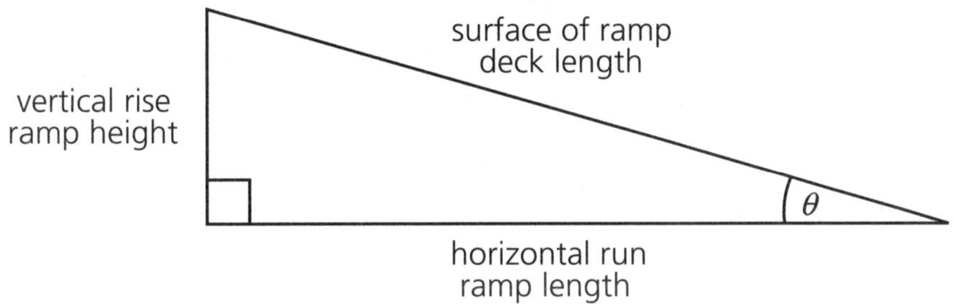

1. According to the law, the slope (the ratio of the vertical rise to the horizontal run) must be no greater than 1:12. This means the maximum ratio of the ramp height to the ramp length is 1:12.

   Complete the following table so the ratio of the height to the length always meets this criterion. Use the Pythagorean theorem to find the deck length.

Height	Length	$\frac{\text{Height}}{\text{Length}}$	Deck Length
3 feet			
	20 feet		
	5 meters		
1 meter			
x meters			
	y feet		
26 inches			

2. Complete the following table in a similar manner. This time, however, use a height to length ratio of 1:15.

Height	Length	Deck Length
36 inches		
	12 feet	
	5 meters	
1.5 meters		
x meters		
	y feet	
24 inches		

Quiz: Right Triangle Trigonometry 2

Name: ______________________

1. You are standing a certain distance from the base of the Washington Monument. The angle of elevation from your feet to the top of the monument is 20°. You know from your tour book that the height of the monument is 555 feet. How far are you from the base of the monument?

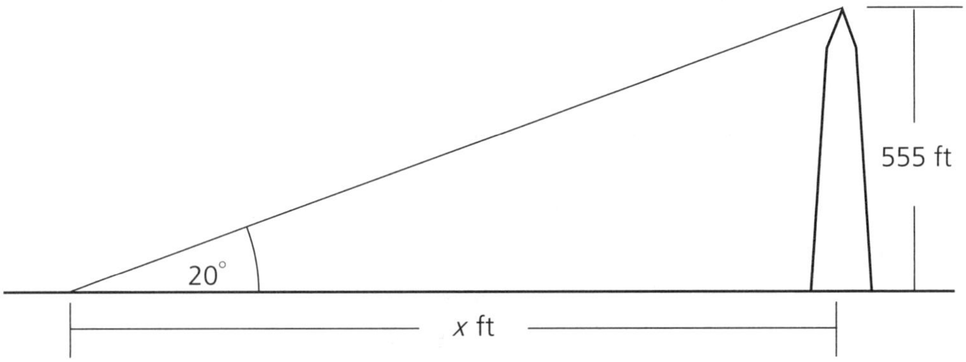

2. A pilot approaches an airport for a landing. The pilot is 50 km from the airport and measures the angle of depression to the airport to be 15°. What is the pilot's altitude?

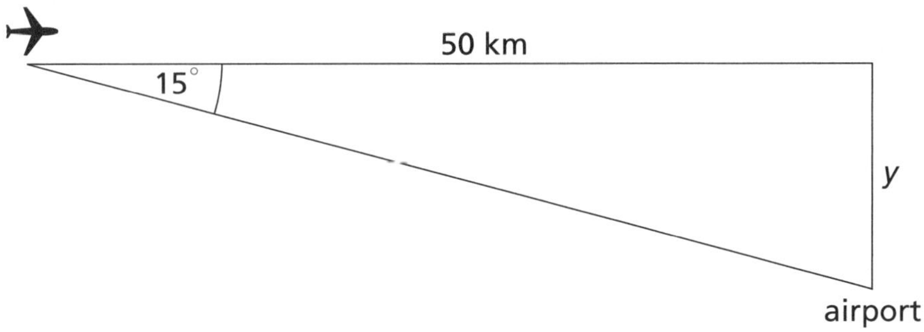

3. A firefighter is called to rescue a cat caught in a tree. If the firefighter is standing 30 m from the base of the tree and if the angle of elevation to the cat is 20°, how high in the tree is the cat?

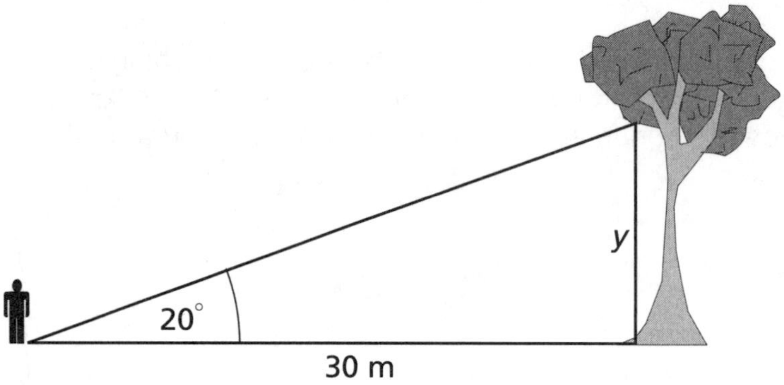

4. You are a camera operator in a blimp over a stadium. Your blimp is 200 m directly over one end zone. The distance between the end zones is 100 m. What angle of depression will allow you to film the action in the *other* end zone?

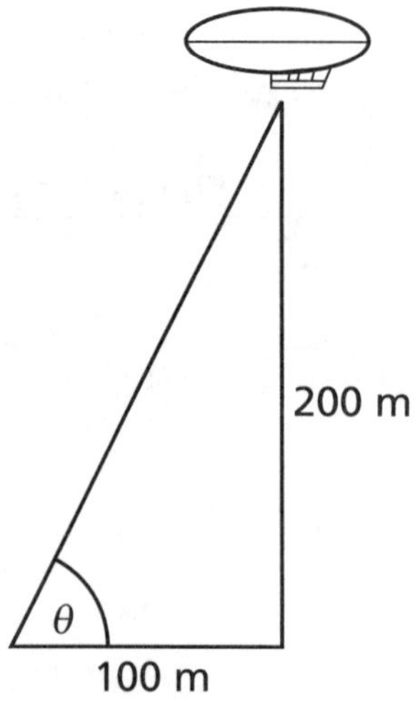

Squares: Part 1

Name: ______________________________

For each equation:

- Enter the equation into your graphing calculator and graph it
- Fill in at least five *x* and *y* values in the table
- Sketch the graph
- Find the *x*-intercepts and the *y*-intercept
- Find the equation for the axis of symmetry
- Find the coordinates of the vertex

1. $y = x^2 + 5x + 4 = (\qquad)(\qquad)$

x	y

x-intercepts: ________________

y-intercept: ________________

Axis of symmetry: *x* = ________

Vertex: (______, ______)

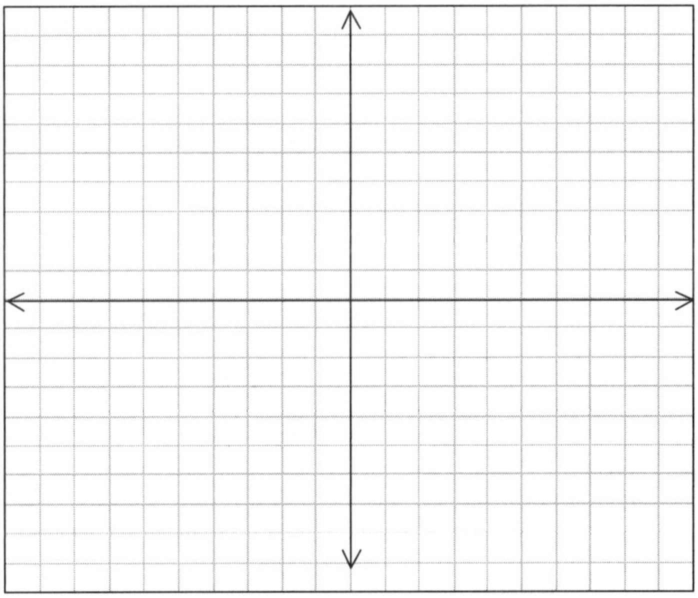

2. $y = x^2 - 5x + 6 = (\quad\quad)(\quad\quad)$

x	y

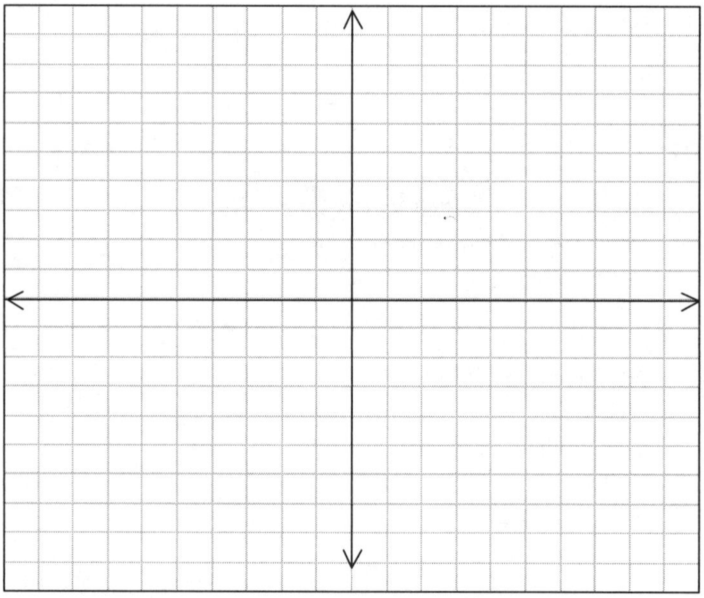

x-intercepts: ______________

y-intercept: ______________

Axis of symmetry: *x* = ________

Vertex: (______, ______)

Squares: Part 2

Name: ______________________

For each equation:

- Enter the equation into your graphing calculator and graph it
- Fill in at least five *x* and *y* values in the table
- Sketch the graph
- Find the *x*-intercepts and the *y*-intercept
- Find the equation for the axis of symmetry
- Find the coordinates of the vertex

1. $y = x^2 + 7x + 6 = (\quad)(\quad)$

x	y

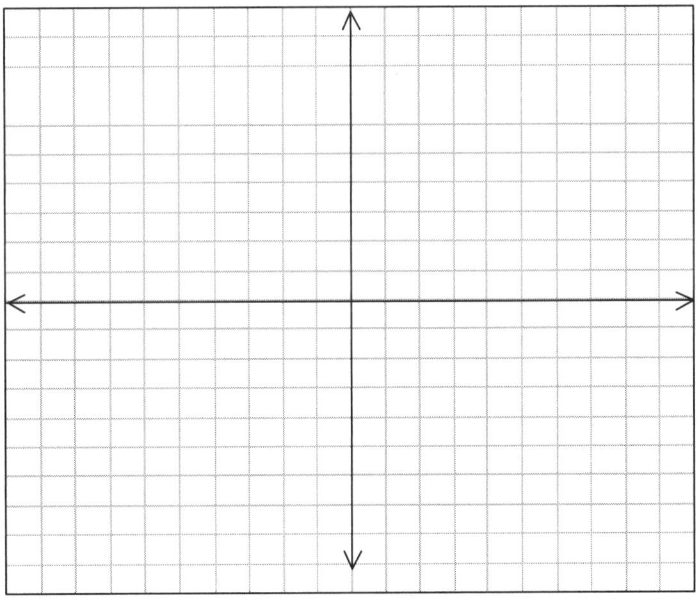

x-intercepts: ______________

y-intercept: ______________

Axis of symmetry: *x* = ________

Vertex: (______, ______)

2. $y = x^2 - 6x + 8 = (\quad\quad)(\quad\quad)$

x	y

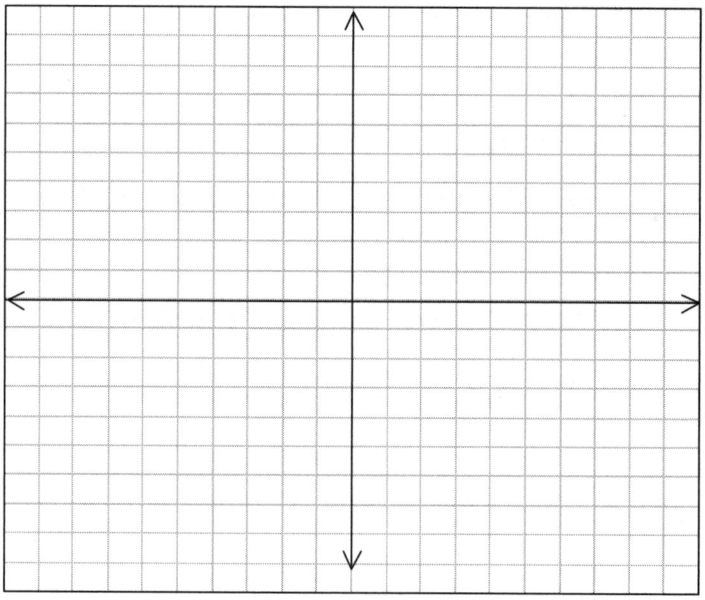

x-intercepts: ______________

y-intercept: ______________

Axis of symmetry: $x =$ _______

Vertex: (_____, _____)

Quadratic Formula

Name: ____________________________

General form of the quadratic equation:

$$y = Ax^2 + Bx + C$$

Quadratic formula:

$$x = \frac{-B \pm \sqrt{B^2 - 4AC}}{2A}$$

For each quadratic equation:

- Write the values of A, B, and C from the general form
- Use the quadratic formula to solve the equation for the x-intercepts
- Use the x-intercepts to write the equation in factored form

1. $y = x^2 - 9x - 10 = (\qquad\qquad)(\qquad\qquad)$

   $A =$ __________ $B =$ __________ $C =$ __________

   x-intercepts: ____________________________________

2. $y = 2x^2 - 16x - 128 = (\qquad\qquad)(\qquad\qquad)$

   $A =$ __________ $B =$ __________ $C =$ __________

   x-intercepts: ____________________________________

3. $y = 3x^2 + 8x - 10 = (\qquad\qquad)(\qquad\qquad)$

   $A =$ __________ $B =$ __________ $C =$ __________

   x-intercepts: ____________________________________

Vertical Motion

Name: ______________________

A cannonball is shot straight up into the air from the side of a mountain. The flight of the cannonball can be modeled by this equation for vertical motion:

$$y = -16x^2 + vx + h$$

In this equation, y is the number of feet above the valley, x is the time in seconds from the moment the ball is shot, v is the initial velocity (or speed) in feet per second, and h is the initial height in feet of the cannonball above the valley floor.

The initial velocity of the cannonball is 600 feet per second, and the cannon is 100 feet above the valley floor.

1. Write the equation for flight of the cannonball.

Use complete sentences to answer the following questions.

2. How high is the cannonball:
   a. 1 second *after* it is shot?

   b. 20 seconds *after*?

   c. 50 seconds *after*?

   d. 5 seconds *before* it is shot?

   e. 50 seconds *before* it is shot?

3. Do all the answers to question 2 make sense in this problem situation?

4. Use the quadratic formula to find when the cannonball is on the ground. How do you know?

5. When is the cannonball at its highest point? How high does it get?

6. Use the quadratic formula to find when the cannonball will be back at its original level. How do you know?

7. When is the cannonball 3000 feet above the valley floor?

8. Use the information you have found so far about the problem situation to fill in this table.

Labels		
Units		
Formula		

9. Graph your equation on the following grid.

Data

Name: ____________________

1. Use the data table to graph the points. Then find the slope, the y-intercept, and the equation that *fits* the data points.

x	y
1	*10*
4	*31*
3	*24*
8	*51*
11	*63*

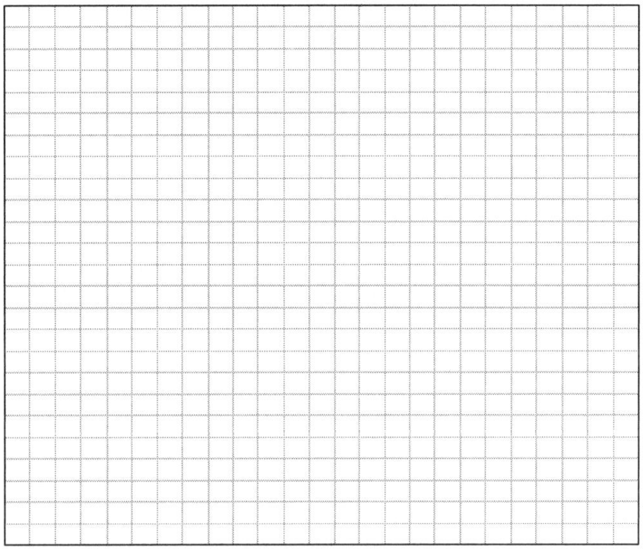

Slope: ____________

y-intercept: ____________

Equation: $y =$ ____________

For each set of data, calculate the mean, the median, and the mode.

2. Data: 14, 23, 24, 19, 15, 18, 19

mean: $\bar{x}$ = ___________ median =_________ mode = _______

3. Data: 1111, 1415, 1612, 1111, 1313

mean: $\bar{x}$ = ____________ median =__________ mode = _______

Running the Mile

Name: ____________________________

This table shows the fastest time in seconds that a human ran the mile from 1945 to 1985. Find the slope, the y-intercept, and the equation of the linear regression line for *number of years since 1945* versus *time in seconds*. Then answer the questions.

Year	***Time in Seconds***
1945	*241.4*
1954	*238.0*
1957	*237.2*
1962	*234.4*
1964	*234.1*
1965	*233.6*
1966	*231.3*
1967	*231.1*
1975	*229.4*
1979	*229.0*
1980	*228.8*
1981	*228.4*
1981	*227.3*
1985	*226.3*

Slope: b = ________________ y-intercept: a = ________________

Equation: y = ________________ Regression coefficient: r = ________________

Write your answers in complete sentences.

1. Use your model (equation) to predict how fast a runner in 1990 should be able to run the mile.

2. Use your model (equation) to predict how fast a runner in 1993 should be able to run the mile.

3. In what year does your model (equation) predict that the fastest time a runner ran the mile was 237.5 seconds?

4. In what year does your model (equation) predict that 200 seconds is the fastest time a runner will run the mile?

5. How good a fit is your model? Explain your thinking.

Probability on the Shuffleboard Court

Name: ______________________

1. Your friend Sasha calls you in the evening while you are on the phone with another friend, Jordan. Sasha is confused about something in his class notes. You decide to conference him so he can also talk to Jordan. Sasha said that he wrote the probability of landing on the trapezoid on the shuffleboard triangle pictured above is 1/3, but without understanding why. He tells you he thinks this is so because on the shuffleboard, he sees three different shapes: a triangle, a rhombus and a trapezoid. Jordan jumps in saying that that reason is wrong since he thinks there are four triangles, a rhombus and a trapezoid there are six shapes, so that the probability of landing on the trapezoid is 1/6, not 1/3. You disagree with Jordan's answer since you have the same answer as Sasha, however you disagree with Sasha's reasoning. Explain your reasoning.

2. How is area related to probability in this problem?

3. Find the probability of landing on the shaded triangle embedded in this square.

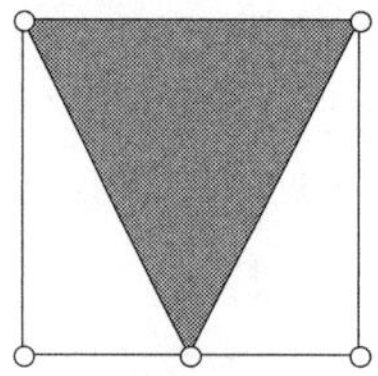

2 cm

Assessments with Answers

Contents

Assessments with Answers

Unit 1: Organizing Single Variable Data

Unit 2: Patterns and Linear Functions

Unit 3: Proportional Reasoning

Unit 4: Modeling Situations Using Multiple Representations

Unit 5: Linear Equations: Solving, Graphing, Intercepts, and Slope

Unit 6: Area and Perimeter

Unit 7: The Geometry and Algebra of Right Triangles

Unit 8: Quadratics

Unit 9: Statistical Analysis

Unit 10: Probability

Baseball

Name:________________________

The data set below represents the top 40 homerun hitters in the American and National Baseball Leagues for 2001.

American League Homerun Leaders 2001	National League Homerun Leaders 2001
1) Alex Rodriguez, Tex 52	1) Barry Bonds, SF 73
2) Jim Thome, Cle 49	2) Sammy Sosa, ChC 64
3) R. Palmeiro, Tex 47	3) Luis Gonzalez, Ari 57
4) Manny Ramirez, Bos 41	4) Shawn Green, LA 49
5) Troy Glaus, Ana 41	5) Todd Helton, Col 49
6) Carlos Delgado, Tor 39	6) Richie Sexson, Mil 45
7) Jason Giambi, Oak 38	7) Phil Nevin, SD 41
8) Bret Boone, Sea 37	8) Jeff Bagwell, Hou 39
9) Juan Gonzalez, Cle 35	9) Chipper Jones, Atl 38
10) Jose Cruz, Tor 34	10) Larry Walker, Col 38
11) Tino Martinez, NYY 34	11) Rich Aurilia, SF 37
12) Paul Konerko, CWS 32	12) Brian Giles, Pit 37
13) Eric Chavez, Oak 32	13) Albert Pujols, StL 37
14) Miguel Tejada, Oak 31	14) Mike Piazza, NYM 36
15) M. Ordonez, CWS 31	15) Gary Sheffield, LA 36
16) Mike Sweeney, KC 29	16) Jeromy Burnitz, Mil 34
17) Jose Valentin, CWS 28	17) Andruw Jones, Atl 34
18) G. Anderson, Ana 28	18) V. Guerrero, Mon 34
19) Ellis Burks, Cle 28	19) Aramis Ramirez, Pit 34
20) Raul Mondesi, Tor 27	20) Lance Berkman, Hou 34
21) Trot Nixon, Bos 27	21) Reggie Sanders, Ari 33
22) Torii Hunter, Min 27	22) Cliff Floyd, Fla 31
23) Jermaine Dye, KC/Oak 26	23) Bobby Abreu, Phi 31
24) Corey Koskie, Min 26	24) Ryan Klesko, SD 30
25) B. Williams, NYY 26	25) Jim Edmonds, StL 30
26) Mike Cameron, Sea 25	26) Mark McGwire, StL 29
27) Tony Batista, Bal/Tor 25	27) J.D. Drew, StL 27
28) Ivan Rodriguez, Tex 25	28) Pat Burrell, Phi 27
29) Carlos Beltran, KC 24	29) Moises Alou, Hou 27
30) Carlos Lee, CWS 24	30) Jose Hernandez, Mil 25
31) Greg Vaughn, TB 24	31) Brian Jordan, Atl 25
32) Ruben Sierra, Tex 23	32) Scott Rolen, Phi 25
33) Edgar Martinez, Sea 23	33) Bubba Trammell, SD 25
34) Jorge Posada, NYY 22	34) Paul Lo Duca, LA 25
35) Brian Daubach, Bos 22	35) Lee Stevens, Mon 25
36) Derek Jeter, NYY 21	36) Vinny Castilla, Hou 23
37) Paul O'Neill, NYY 21	37) Preston Wilson, Fla 23
38) John Olerud, Sea 21	38) Jeff Kent, SF 22
39) Ray Durham, CWS 20	39) K. Griffey Jr., Cin 22
40) Marty Cordova, Cle 20	40) Dmitri Young, Cin 21

Source: www.espn.com

1. To compare the two leagues construct
   a. Stem and Leaf Plot
   b. Histogram

National League Stem and Leaf Plot

Stem	Leaves
2	***1 2 2 3 3 5 5 5 5 5 5 7 7 7 9***
3	***0 0 1 1 3 4 4 4 4 4 6 6 7 7 7 8 8 9***
4	***1 5 9 9***
5	***6***
6	***4***
7	***3***
	stem width = 10

America League Stem and Leaf Plot

Stem	Leaves
2	***0 0 1 1 1 2 2 3 3 4 4 4 5 5 5 6 6 6 7 7 7 8 8 8 9***
3	***1 1 2 2 4 4 5 7 8 9***
4	***1 1 7 9***
5	***2***
6	
7	
	stem width = 10

National League Histogram

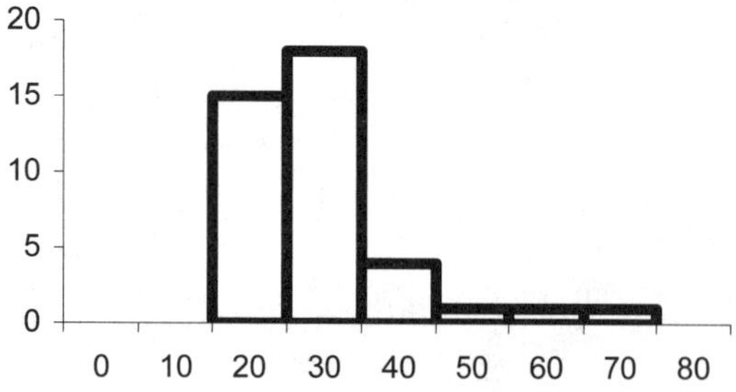

American League Histogram

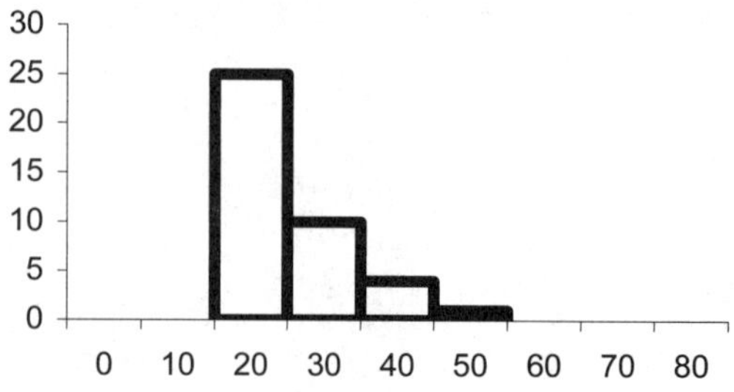

2. What type of information can you gather from each type of plot?

From the stem and leaf plot you can determine the distribution of the data and therefore which league hit more homeruns. You can also see specifically the number of homeruns hit per person in the data set. From the histogram you can also see a distribution, but it will be the frequency of a homeruns hits within an interval.

3. Based on the stem and leaf plots what can you say about the performance of the two leagues?

Based on the stem and leaf plot, the distribution is skewed to the right, meaning the majority of the 40 top homerun hitters hits between 20 and 40 homeruns in each league. The national league had more people hit homeruns between 50 and 80. The American League had most of its top 40 hitters hit between 20 and 30 homeruns.

4. Based on the histogram what can you say about the performance of the two leagues?

Based on the histogram, you can see that the American League had the highest frequency of homeruns hit in the interval 20-24 and 25-29. The performance is shown to be almost equivalent, but the National League has a higher frequency of homeruns in the intervals from 50 to 70.

5. Do the two displays provide the same information? Explain your reasoning.

Students may say yes, since it is the same data set, however you want the answer to reflect the fact that the stem and leaf allows you to see actual numbers while the histogram only provides the frequency of individuals who hit homeruns within an interval.

6. Which display is more informative? Why?

Answers will vary, but students should focus on stating the original purpose for generating a display, such as wanting a one-for-one view of the number of homeruns hit by the top forty players. Then the stem and leaf would be the appropriate answer.

7. From which distribution is it easier to determine the mean number of home runs?

Since the stem and leaf give exact values it will be easier to use to determine the mean.

Number Patterns Test

Name: ______________________________

Task 1

For each sequence, find the next 3 terms and describe the pattern.

1. 3, 8, 13, 18, 23, <u>28</u>, <u>33</u>, <u>38</u>..

   Add five to each term. OR Multiply the term number by 5 and subtract 2.

2. 1, 2, 3, 5, 8, 13, ***21, 34, 55,***

   Add two previous terms to get the next term.

3. 1, 4, 9, 16, ***25, 36, 49***,...

   Each term is the term number squared. OR Add the next odd number.

4. 1, 100, 2, 99, 3, 98, ***4***, ***97***, ***5***,...

   Odd numbered terms increase by 1. Even numbered terms decrease by 1.

5. 3, -9, 27, -81, ***243***, ***-729***, ***2187***,...

   Multiply the previous term by –3 each time. OR $(-1)^{n-1}(2)^n$

Task 2

For each sequence, find the 10th term and describe the pattern.

1.

Term Number	*1*	*2*	*3*	*4*	*...*	*10*	
Sequence	*5*	*15*	*25*	*35*	*...*	***95***	

The function is 10n – 5, so multiply the term number by 10 and subtract 5.

2.

Term Number	*1*	*2*	*3*	*4*	*...*	*10*	
Sequence	*2*	*-4*	*8*	*-16*	*...*	***-1024***	

2^n***, even terms negative, odd terms positive OR*** $(-1)^{n-1}(2)^n$ ***OR Multiply the previous term by 2.***

3.

Term Number	*1*	*2*	*3*	*4*	*5*	...	*10*
Sequence	*1*	*5*	*11*	*55*	*111*	...	***55555***

Odd term N length string of 1's, even terms N/2 length string of 5's

Task 3

For each sequence, find an expression for the Nth term and describe the pattern.

1.

Term Number	*1*	*2*	*3*	*4*	...	*N*	
Sequence	*2*	*7*	*12*	*17*	...	***5n-3***	

Multiply the term number by 5 and subtract 3.

2.

Term Number	*1*	*2*	*3*	*4*	...	*N*	
Sequence	*5*	*25*	*125*	*625*	...	5^n	

Raise five to the power of the term number.

3.

Term Number	*1*	*2*	*3*	*4*	...	*N*	
Sequence	*10*	*18*	*26*	*34*	...	***8n + 2***	

Multiply each term number by 8 and add 2 to find each term.

Toothpick Test

Name: ____________________________

The following toothpick diagrams show a single square constructed from 4 toothpicks and two squares constructed from 7 toothpicks.

One Square

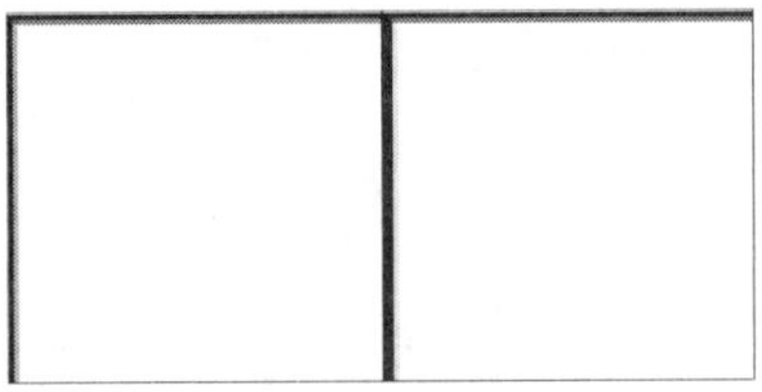

Two Squares

1. Draw toothpick diagrams that show how to construct three squares, four squares, and five squares in the same manner.

   a. Three squares:

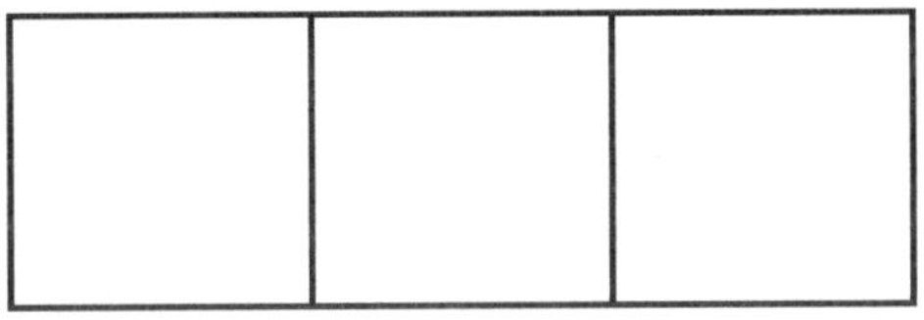

   b. Four squares:

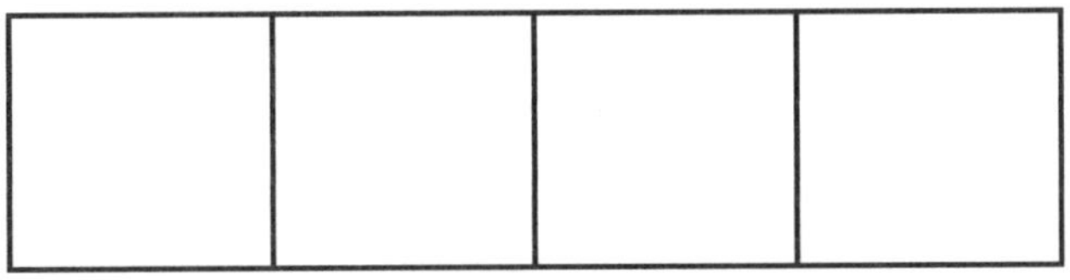

   c. Five squares:

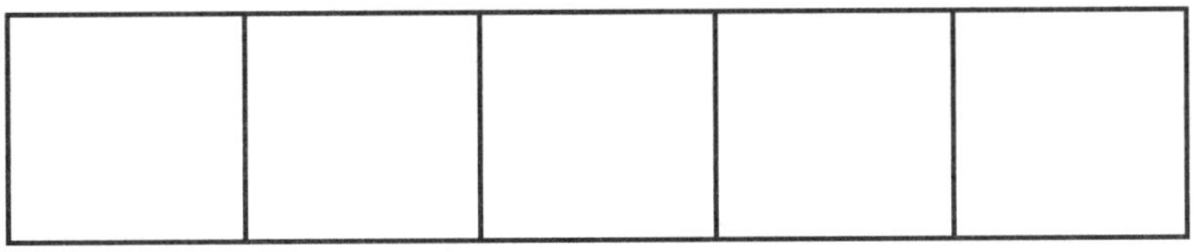

2. Complete this table.

	Number of Squares	Number of Toothpicks
Units	***Squares***	***Tooth Picks***
	1	***4***
	2	***7***
	3	***10***
	4	***13***
	5	***16***
	6	***19***
	10	***31***
	50	***151***

3. Use the information in the table and your diagrams to describe the pattern produced by constructing the toothpick squares. Write your answer in complete sentences.

Multiply the number of desired squares by 3 and add 1.

4. Find an expression that gives the number of toothpicks required to construct *N* squares.

3n + 1

Earning Money

Name: ____________________________

A friend of yours is employed at a company where he makes $5.25 an hour.

Make a table that shows his earnings for various numbers of hours worked (between 1 hour and 20 hours). Include the algebraic expression in your table. Use this information to construct a line graph.

Labels	***Time***	***Earnings***
Units	***hours***	***dollars***
Formula	***x***	***5.25x***
	1	***5.25***
	2	***10.50***
	3	***15.75***
	5	***26.25***
	10	***52.50***
	15	***78.75***
	20	***105***

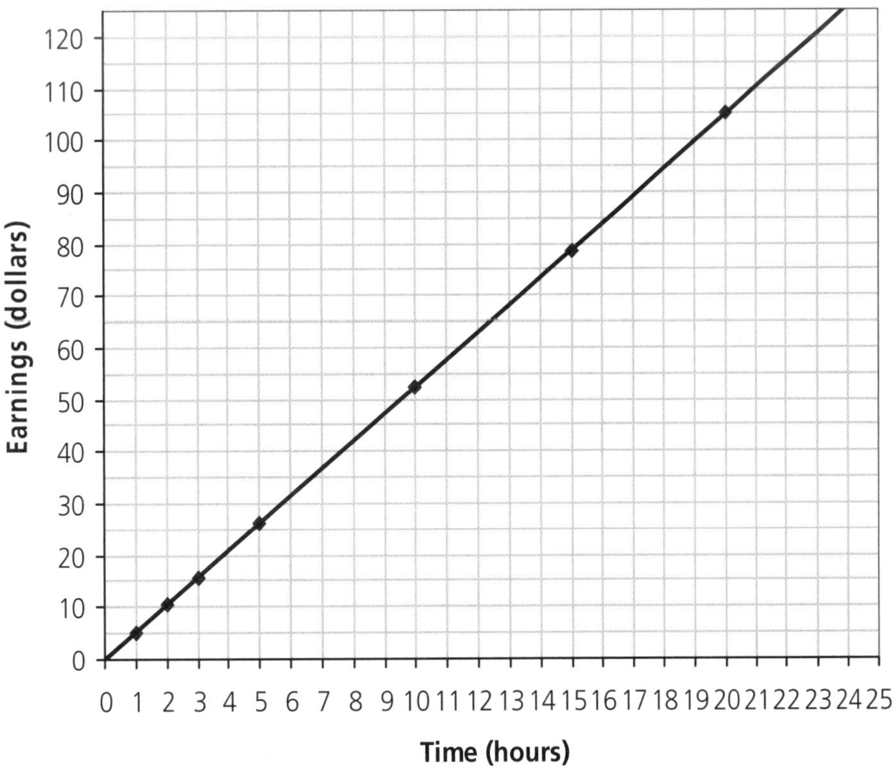

Vacation Flights

Name: ____________________________

You are going on vacation by airplane. The average speed of the plane is 325 miles per hour.

Make a table that shows how far you will travel for various times between 1 hour and 20 hours. Include the algebraic expression in your table. Use this information to construct a line graph.

Labels	***Time in the Air***	***Distance Traveled***
Units	***Hours***	***Miles***
Expression	***x***	***325x***
	1	***325***
	2	***650***
	3	***975***
	5	***1625***
	10	***3250***
	20	***6500***

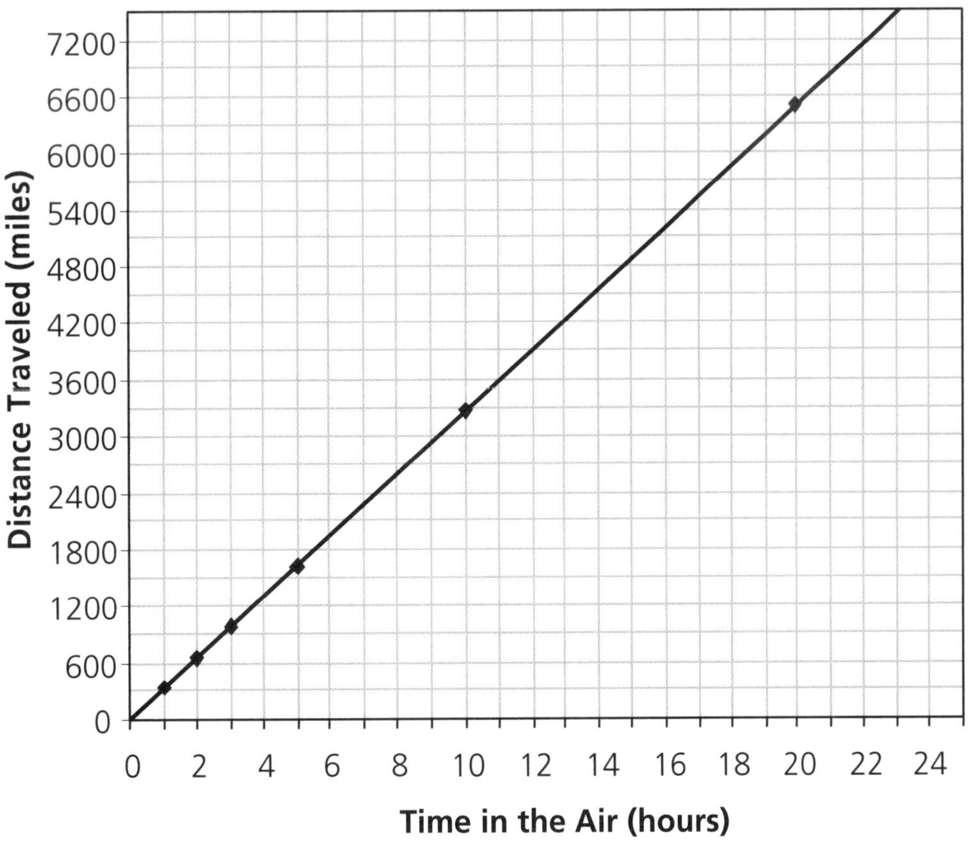

Students Who Don't Take Algebra

Name: ____________________________

One out of every four high school students never takes algebra.

Make a table that shows how many students don't take algebra at schools of various sizes. Include the algebraic expression in your table. Use this information to construct a line graph.

Labels	***High School Students***	***High School Students who do not take Algebra***
Units	***Students***	***Students***
Expression	***s***	***(1/4)x***
	0	***0***
	50	***12.5***
	100	***25***
	150	***37.5***
	200	***50***
	250	***62.5***
	300	***75***

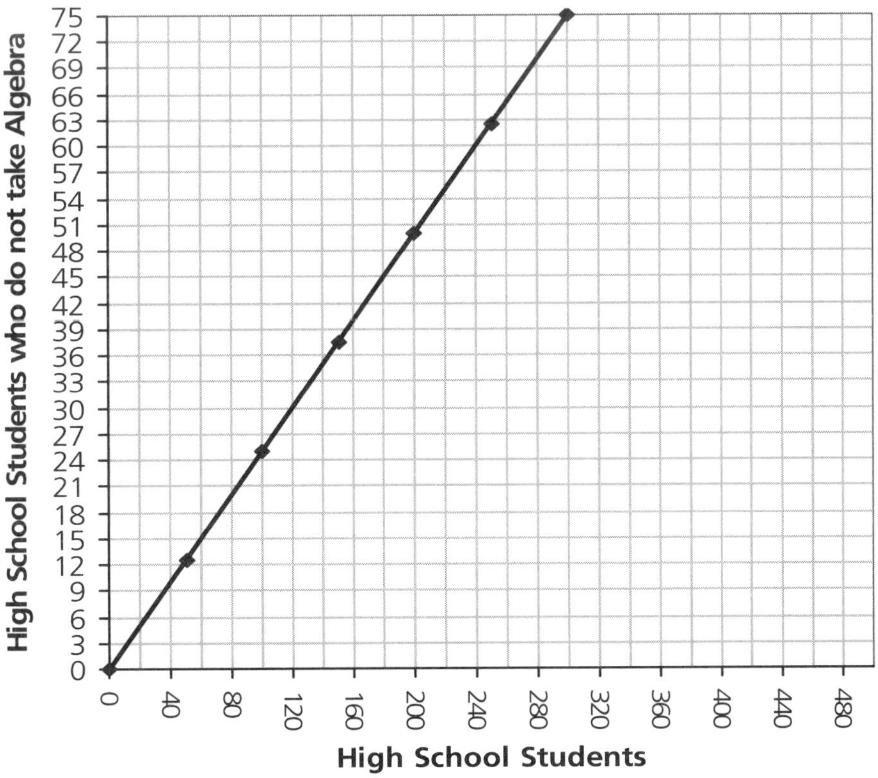

Paying Taxes

Name: ______________________________

The average family in the United States pays about 38% of their income in taxes of all kinds.

Make a table that shows the amount of taxes paid by families with various incomes. Include the algebraic expression in your table. Use this information to construct a line graph.

Labels	***Income***	***Taxes Paid***
Units	***$***	***$***
Expression	***x***	***0.38x***
	0	***0***
	5000	***1900***
	10,000	***3800***
	15,000	***5700***
	20,000	***7600***
	25,000	***9500***
	30,000	***11,400***

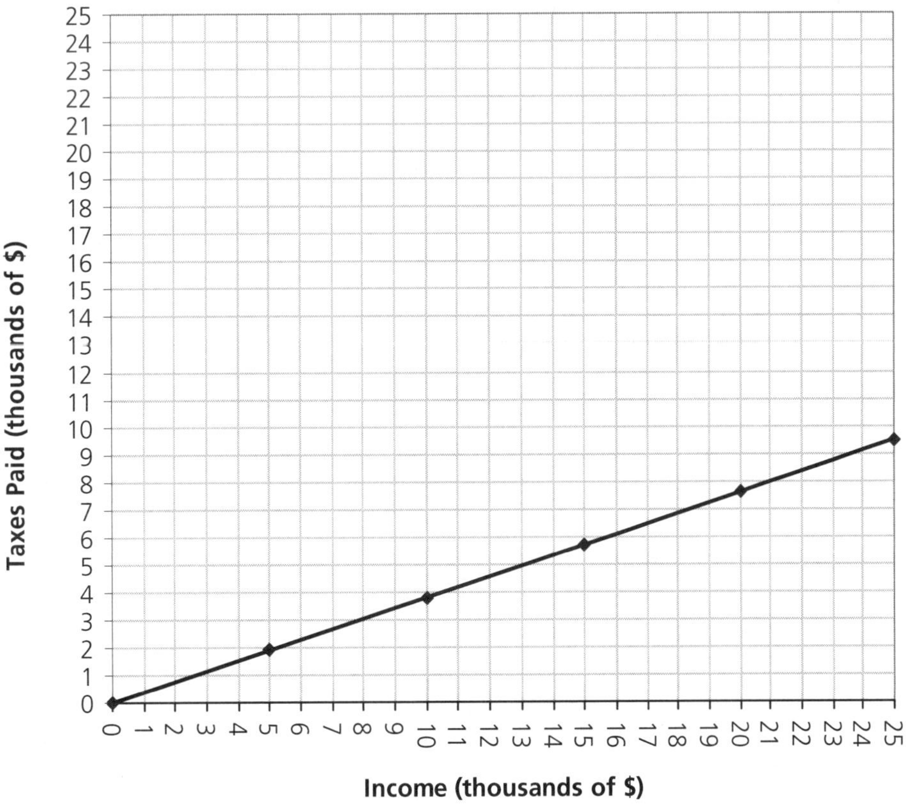

Stocking the Shelves

Name: ______________________________

You are a volunteer for the school store. One of the most popular items is cool ranch flavored tortilla chips. There are two local vendors that will delivery the chips to the school store. Each makes their deliveries on the first of every month. Chips-R-Us charges 39 cents per bag with a monthly delivery fee of $25. It's In The Bag charges 18 cents per bag with a monthly delivery fee of $65.

1. Write an algebraic equation to model the monthly cost of purchasing chips from Chips-R-Us.

   y = 0.39x + 25

   x = # of bags of chips purchased

   y = Cost of purchasing from Chips-R-Us

2. Write an algebraic equation to model the monthly cost of purchasing chips from It's In The Bag.

   y = 0.18x + 65

   x = # of bags of chips purchased

   y = Cost of purchasing from It's In The Bag

3. Complete the following table summarizing the cost of purchasing chips from each vendor based upon the actual sales by month for the last school year.

	Month	Number of bags of chips purchased	Cost of purchasing from Chips-R-Us	Cost of purchasing from It's In The Bag
Label	**Month**	**Number of bags of chips purchased**	**Cost of purchasing from Chips-R-Us**	**Cost of purchasing from It's In The Bag**
Units		***Bags***	***Dollars***	***Dollars***
	September	187	***97.93***	***98.66***
	October	229	***114.31***	***106.22***
	November	162	***88.18***	***94.16***
	December	137	***78.43***	***89.66***
	January	171	***91.69***	***95.78***
	February	201	***103.39***	***101.18***
	March	192	***99.88***	***99.56***
	April	258	***125.62***	***111.44***
	May	210	***106.90***	***102.80***
	June	79	***55.81***	***79.22***
Expression		***x***	***0.39x+25***	***0.18x+65***

4. Generate a graph displaying the cost of purchasing from both Chips-R-Us and It's In The Bag.

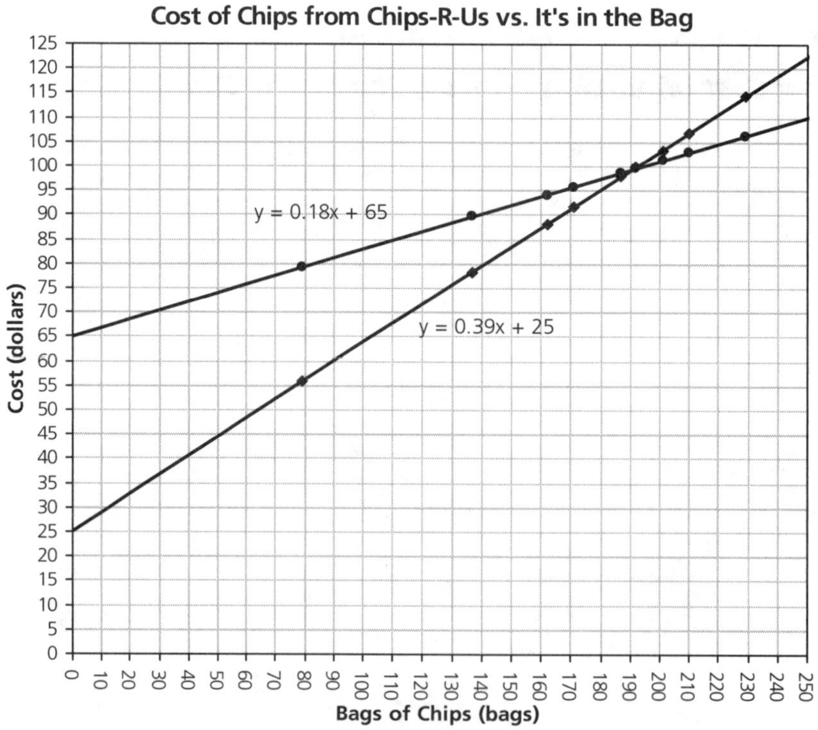

5. How many bags of chips can you purchase with each vendor if you have the following budgets? Explain how you determined each and write your answer in a complete sentence.

   a. $50

Chips-R-Us	***64 bags***
It's In The Bag	***0 bags***

   b. $125

Chips-R-Us	***256 bags***
It's In The Bag	***333 bags***

   c. $200

Chips-R-Us	***448 bags***
It's In The Bag	***750 bags***

6. Compare the price to purchase from each of the two vendors. Prepare a written analysis indicating which vendor will cost the least and when. Be sure to include when the costs will be the same.

   There is an equal cost of $99.29, when approximately 190 bags are purchased. When less than 190 bags are purchased, Chips-R-Us is less expensive. When more than 190 bags are purchased, It's In The Bag is less.

7. If you had to choose one vendor for the entire school year, which would you choose? Would your choice change if you could switch vendors each month? Explain your choices using complete sentences.

   Answers will vary.

Integers and Solving Equations: Form A

Name: ______________________________

Find the answer to each problem without using your calculator.

1. $7 + (-14) =$ ***-7***
2. $-19 + 110 =$ ***91***
3. $-2 \bullet 10 =$ ***20***
4. $\frac{210}{-10} =$ ***-21***
5. $13 - 25 =$ ***-12***
6. $-10(-6) =$ ***60***
7. $-23 + (-14) =$ ***-37***
8. $\frac{-41 + 33}{-2} =$***4***
9. $\frac{10 + (-4) - 4}{2} =$***1***
10. $20(-3) - 10 =$ ***-70***

Solve each equation. Show all your work.

11\. e ı 5 = 8 ***e = 8 – 5 = 3***

12\. *f* ı 9 = 13 ***f = 13 –9 = 4***

13\. $-2g = -20$ ***g=-20/-2 = 10***

14\. $3h = 18$ ***h= 18/3= 6***

15\. $20x = -10$ ***x = -10/20 = -1/2***

16\. $5j = -25$ ***j= -25/5 = -5***

17\. $7 = \frac{k}{-12}$; ***k = 7(-12) = -84***

18\. $-2 - L = 18$ ***L= -2-18=-20***

Integers and Solving Equations: Form B

Name: ______________________

Find the answer to each problem without using your calculator.

1. 17 + (-14) = ***3***

2. -190 + 110 = ***-80***

3. -21 • 10 = ***-210***

4. $\frac{-50}{-10}$ = ***5***

5. -13 – 5 =***-18***

6. -10(-16) =***160***

7. -213 + (-124) = ***-337***

8. $\frac{-1-3}{-2}$ = ***2***

9. $\frac{10-4-4}{-2}$ = ***-1***

10. 2(-3) – 10 = ***-16***

Solve each equation. Show all your work.

11. 2*E* + 6 = 10 ***2E = 4*** ***E = 4/2 = 2***

12. 4*x* + 8 = 32; ***4x = 24 x = 24/4 = 6***

13. -2*G* + 26 = 50 ***-2G = 24*** ***G = 24/-2 = -12***

14. 3*h* – 18 = 18 ***3h = 36 h =36/3=12***

15. 22*x* + 11 = -11 ***22x = -22*** ***x= -22/22 = -1***

16. 3*J* + (-6) = -15 ***3j = -9 j = -9/3 = -3***

17. $12 = 12 + \frac{k}{-7}$ ***$0 = \frac{k}{-7}$*** ***k = 0***

18. -4 – 4*L* = 16 ***-4L = 20*** ***L =-5***

At the Driving Range

Name: ____________________________

Inspired by the saying "practice makes perfect, "you decide to perfect your golf swing by going to the local driving range. Unfortunately, you do not have your own set of clubs yet so you must rent from the range. Each bucket of 100 golf balls costs $7.50 and it costs $3.00 to rent clubs for 2 hours.

1. Write an algebraic equation for the total cost at the driving range dependent on the number of buckets of balls you buy. Write your equation using functional notation with C(x) representing the cost.

   C(x) = 7.50x + 3

   X = # of buckets of balls

   C(x) = Total cost

2. Show all work and answer in complete sentence

   a. What is the total cost to hit five buckets worth of balls?

      $40.50

   b. How many buckets can you buy with $20?

      2 buckets

   c. Find x if C(x) = 28.

      3.33

   d. Find C(10).

      78

3. Generate a graph displaying the total cost of practicing at the driving range

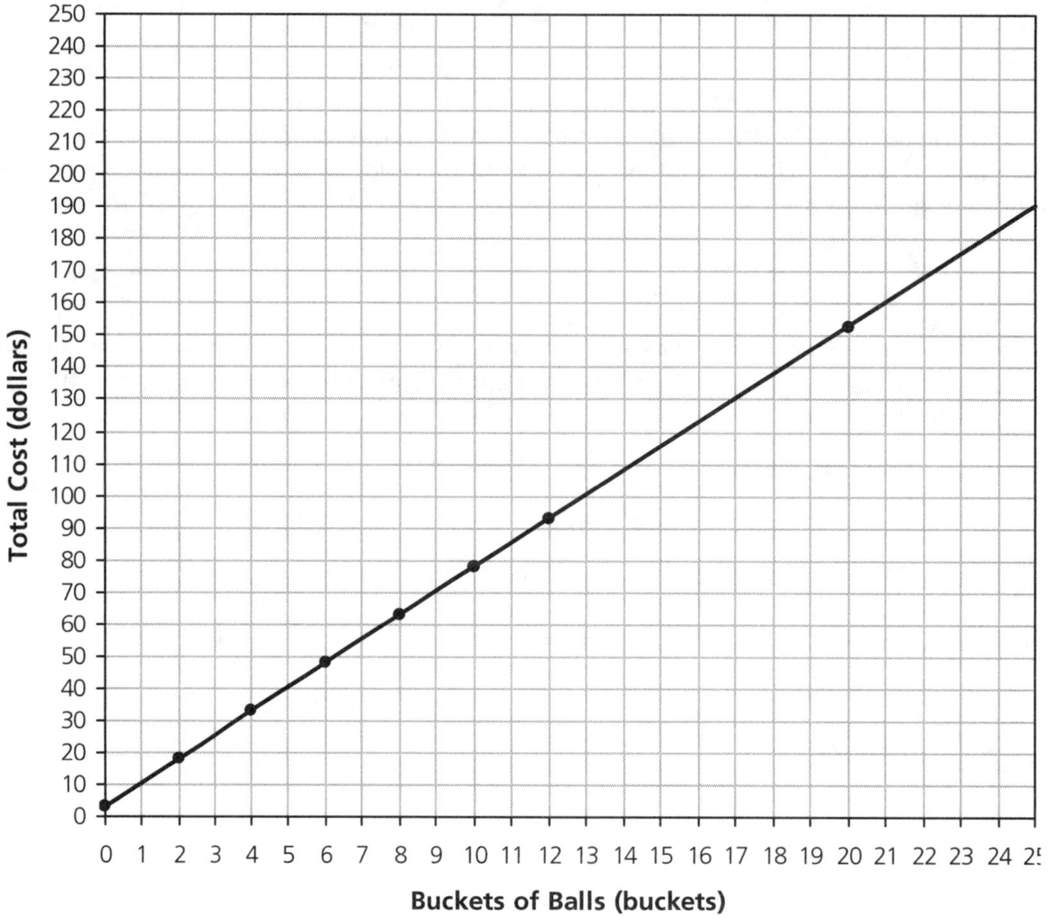

4. The owner of the driving range is looking at the receipts for the day. Are the following costs correct? Explain why or why not in a complete sentence.

   a. $48 for 6 buckets and club rental

   Yes, based on the cost of a bucket of balls.

   b. $70 for 9 buckets and club rental

   No, since the cost would be $72 based on the cost of a bucket of balls.

5. What would be the domain based on the problem situation?

    Since club rental is only for two hours domain should consist of 0 buckets to the maximum number of buckets you could hit in two hours.

6. Based on your answer to question 5, what would be the range based on the problem situation?

    Range could both start at 0 or 3 and end with C(x) for the maximum value in the domain.

7. Set up an inequality to solve each. Graph the solution to each.

    a. Determine the number of buckets of balls you could hit if you had to spend at most $50 but at least $25

    You could hit between 2 and 6 buckets

    b. C(x) > 55 or C(x) <= 40

    x > 6.93 OR x <= 4.93

Balloon Problem

Name: ______________________

Here you will examine a problem situation about buying and selling helium balloons. Show all your work and write your answers in complete sentences.

You have the opportunity to run the helium balloon concession at a local fair. In order to sell balloons, you must pay a $25 fee to the fair sponsors and buy $75 worth of balloons and helium. You can sell each balloon for $0.80. You need to calculate your net profit for the fair. Remember, profit is the money you bring in minus the expenses you have to pay out.

1. How much profit will you have if you sell:

   a. 100 balloons? ***-$20***

   b. 200 balloons? ***$60***

   c. 500 balloons? ***$300***

   d. 50 balloons? -***$60***

   e. 80 balloons? -***$36***

   Write a complete sentence describing how you found these answers.

   Multiply the number of balloons by 0.80 and subtract 100.

2. Do your answers in 1d and 1e *make sense*? Why or why not?

   Yes, these answers make sense because negative numbers for profit indicate that your expenses are greater than your income.

3. Algebraic Expression or Equation for Profit ___***0.80x - 100***__

4. How many balloons must you sell to make a *profit* of:

   a. \$40? ***175 balloons***

   b. \$100? ***250 balloons***

   Write a complete sentence describing how you found these answers.

   Add 100 and then divide by 0.80.

5. How many balloons must you sell to post the following *losses?* (Note: A loss is a negative number. For example, a loss of \$20 is -\$20.)

   a. \$20 ***100 balloons***

   b. \$50 ***62.5 balloons***

   c. \$100 ***0 balloons***

   Write a complete sentence describing how you found these answers.

   Add 100 and then divide by 0.80.

6. Use the information you have found so far to complete this table.

Labels	**Balloons Sold**	**Profit**
Units	***Balloons***	***$***
Expression	***x***	***0.80x - 100***
	200	***60***
	500	***300***
	50	***-60***
	225	*80*
	175	*40*
	250	*100*
	100	*-20*
	10	***-92***
	0	***-100***
	300	***140***
	400	***220***
	275	*120*
	312.5	*150*
	120	*-4*

7. Use the information from the table to construct a graph that illustrates this problem situation.

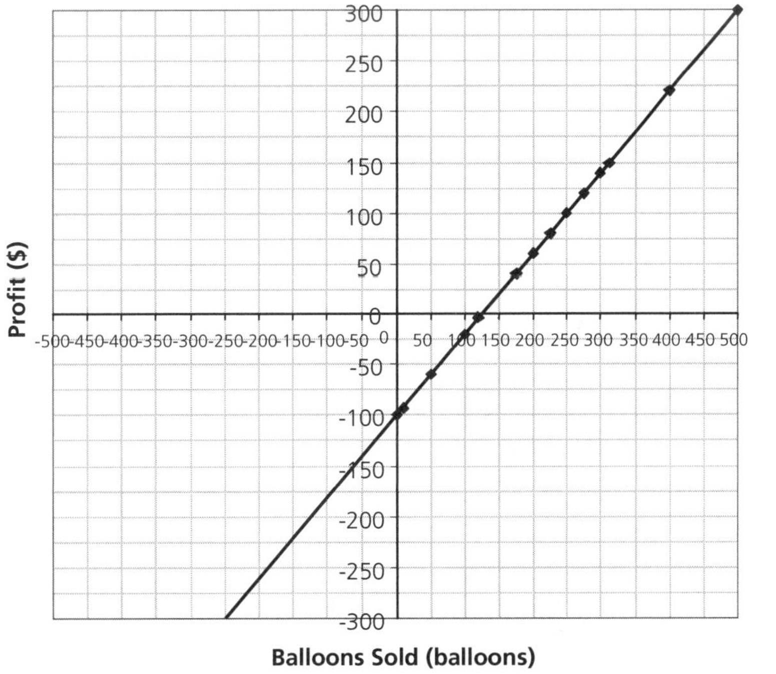

8. How many balloons do you have to sell in order to *break even*? Justify your answer, and write a complete sentence describing how you found this answer.

   You need to sell 125 balloons to break even. This can be seen by looking at the graph. The x-intercept is 125. When you sell 125 balloons, you will have covered all your investments and are ready to begin to make money.

Television Problem and Climber Problem

Name: ____________________________

Television Problem

Use this graph to answer the questions about televisions.

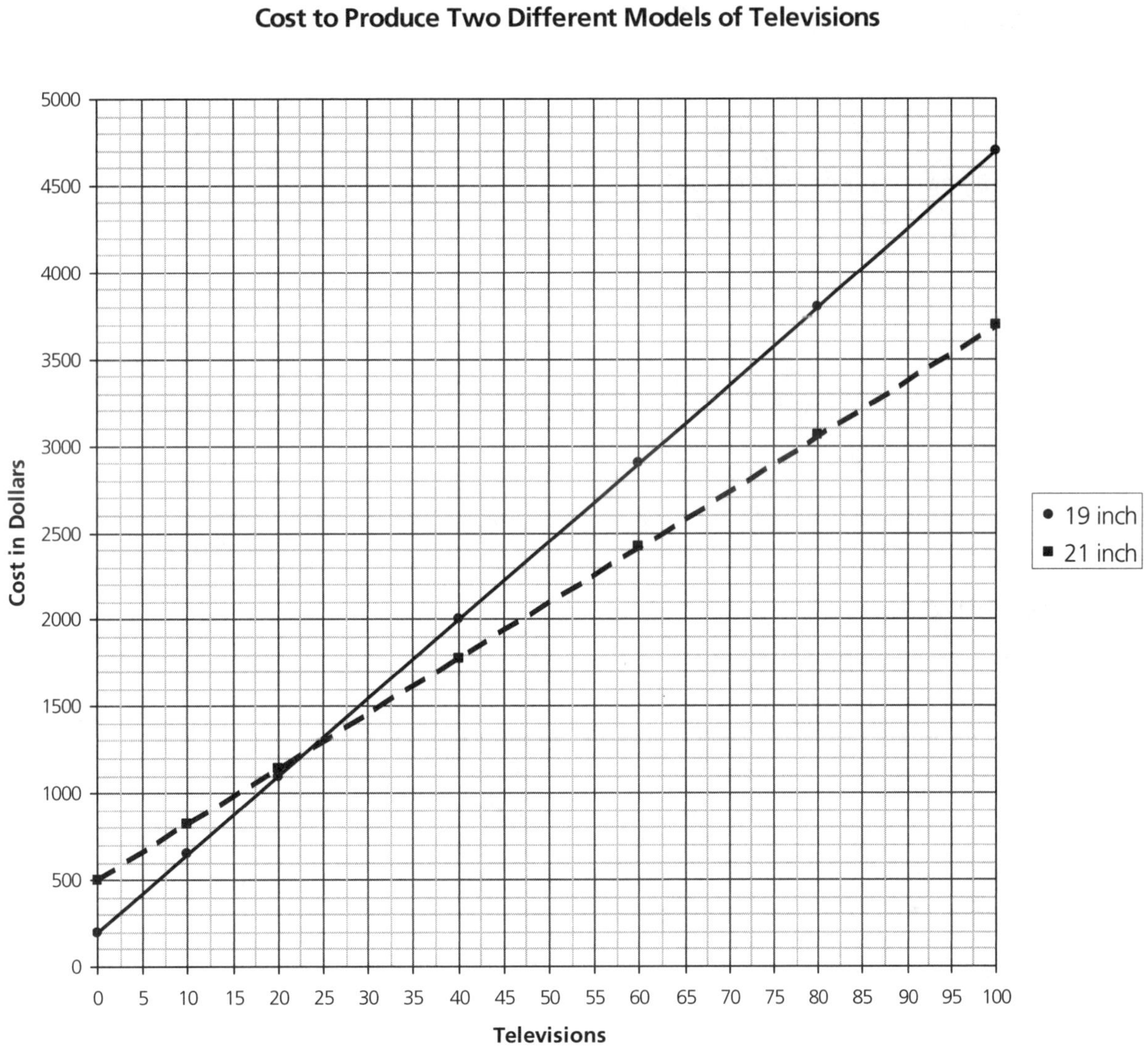

1. What information does this graph show?

   The cost to produce 19 in. and 21 in. TVs.

2. How much does it cost to produce:
   a. thirty-five 19-inch televisions?

      $1800

   b. twenty-five 19-inch televisions?

      $1350

   c. thirteen 19-inch televisions?

      $800

3. How much does it cost to produce:
   a. forty-five 21-inch televisions?

      $1950

   b. fifteen 21-inch televisions?

      $1000

   c. eight 21-inch televisions?

      $750

4. How many of each type of television must be produced for the costs to be the same? How do you know?

   $750

5. Use the graph to predict the cost of producing:
   a. sixty 19-inch televisions.

      $2900

   b. sixty 21-inch televisions.

      $2420

   c. one hundred 21-inch televisions.

      $3700

   d. one hundred 19-inch televisions.

      $4700

Extra credit:

Find an algebraic expression or equation for the cost to produce 19-inch televisions. Find another algebraic expression or equation for the cost to produce 21-inch televisions.

Rule for producing 19-inch televisions: ____***45x+200***____________

Rule for producing 21-inch televisions: ____***32x+500***____________

Climber Problem

A mountain climber is currently 77 feet above ground level on the side of a cliff. She is climbing at the rate of about 7 feet per minute.

Answer the following questions in complete sentences.

1. How many feet above ground level will she be in:

   a. 10 minutes? ***147 ft***

   b. 25 minutes? ***252 ft***

2. Currently, how many feet above ground level is she?

   77 ft

3. How many feet above ground level was she:

   a. 5 minutes *ago*? ***42 ft***

   b. 10 minutes *ago*? ***7 ft***

   c. 15 minutes *ago*? ***-28 ft***

4. How many minutes *ago* did she start her climb?

   11 minutes

5. When will she be:

   a. 154 feet above ground level?

      11 minutes

   b. 175 feet above ground level?

      14 minutes

   c. 287 feet above ground level?

      30 minutes

6. The top of the cliff is 450 feet above ground level. When will she reach the top of the cliff?

   53.29 minutes

7. Write an algebraic rule for the climber's height above ground level.

   y = 7x + 77

   x = time in minute

   y = height in feet

8. Fill in this table and construct a graph for this problem.

Labels	Time	Height
Units	Minutes	Feet
Expression	x	7x + 77
	10	147
	25	252
	0	77
	-5	42
	-10	7
	-11	0
	11	154
	14	175

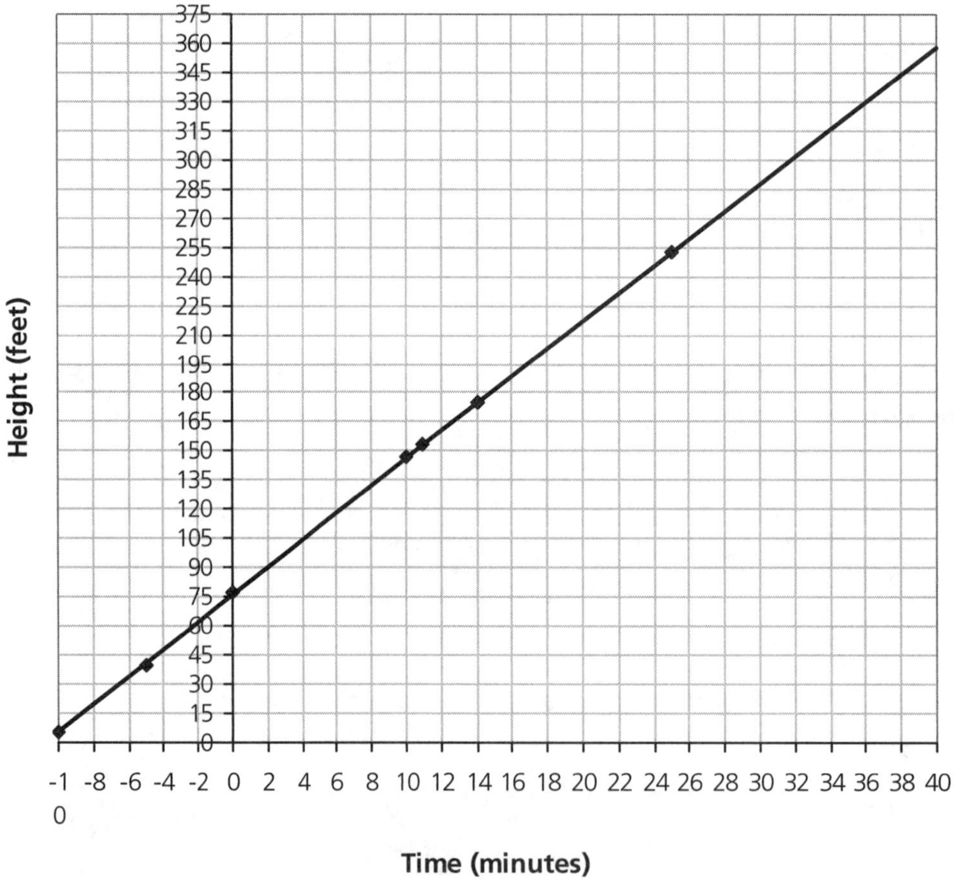

Widgets-R-Us

Name: ____________________________

Widgets cost $4 each, plus $10 per order for shipping and handling.

For this problem situation, find an appropriate algebraic equation, define each variable, make a table of values, construct a graph, and use this information to answer the questions.

1. Find an algebraic equation for this situation.

   Equation: y = ***4x + 10***

   Define each variable in your equation by writing a short phrase that describes what the variable represents.

   x = ***number of widgets***

   y = ***total cost***

2. Make a table with at least 5 values of *x* and *y*.

x	y
2	***18***
5	***30***
-2	***2***
-5	***-10***
0	***10***

3. Graph the equation for this problem situation.

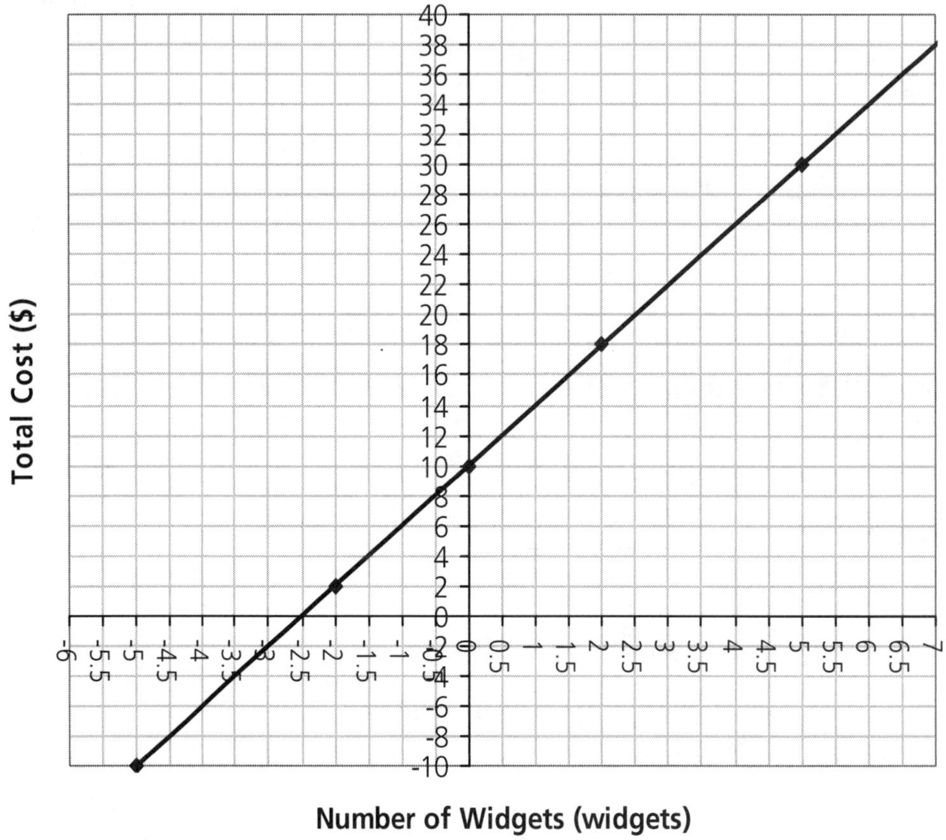

Use your graph, your equation, or your table to answer the following questions. Write your answers in complete sentences.

4. How would you describe your graph?

Answers will vary.

5. Where does the graph intersect the *x*-axis? Why?

The graph intersects the x-axis at x = -2.5 because when y = 0, x = 2.5.

6. Where does the graph intersect the *y*-axis? Why?

The graph intersects the y-axis at y = 10 because when x = 0, y = 10.

7. What is the cost if you order:

   a. 7 widgets? ***$38***

   b. -9 widgets? ***-$26***

8. How many widgets did you order if the total cost is:

   a. $66? ***14 widgets***

   b. -$10? ***-5 widgets***

9. Write an equation that states: the cost of an order of widgets is $212. Solve this equation to find how many widgets you ordered.

   4x + 10 = 212

   4x = 202

   x = 50.5 widgets

10. Find the cost if you order -23,456 widgets.

    -$93,814

11. Do problems 9 and 10 *make sense* in this problem situation? Why or why not?

    No, you must order a whole number of widgets and the cost and number of widgets must be positive.

Computer Rental Problem

Name: ____________________________

For a college computer class in which you are enrolled, you must rent a computer. Two rental companies carry the particular model you need.

Computronics rents the computer you need for $2.50 a day plus a one-time registration fee of $27. A second company, Computers for Students, rents the same computer at a rate of $4.75 per day.

You know you will need the computer for 30 days or less, but you don't know exactly how many days. From whom should you rent? In order to answer this question, write the problem situation for each company, construct a table of values, and find the algebraic expression. Then draw graphs for both companies on a single grid.

Task 1

Write the problem situation for Computronics. Construct a table of values and find the expression.

Problem Situation:

Labels	***Time***	***Cost***
Units	***Days***	***$***
Expression	***t***	***2.50t+27***
	0	***27.00***
	5	***39.50***
	10	***52.00***
	20	***77.00***
	30	***102.00***

Task 2

Write the problem situation for Computers for Students. Construct a table of values and find the expression.

Problem Situation:

Labels	***Time***	***Cost***
Units	***Days***	***$***
Expression	***t***	***4.75t***
	0	***0***
	5	***23.75***
	10	***47.50***
	20	***95.00***
	30	***142.50***

Task 3

Draw graphs that show the information from each company.

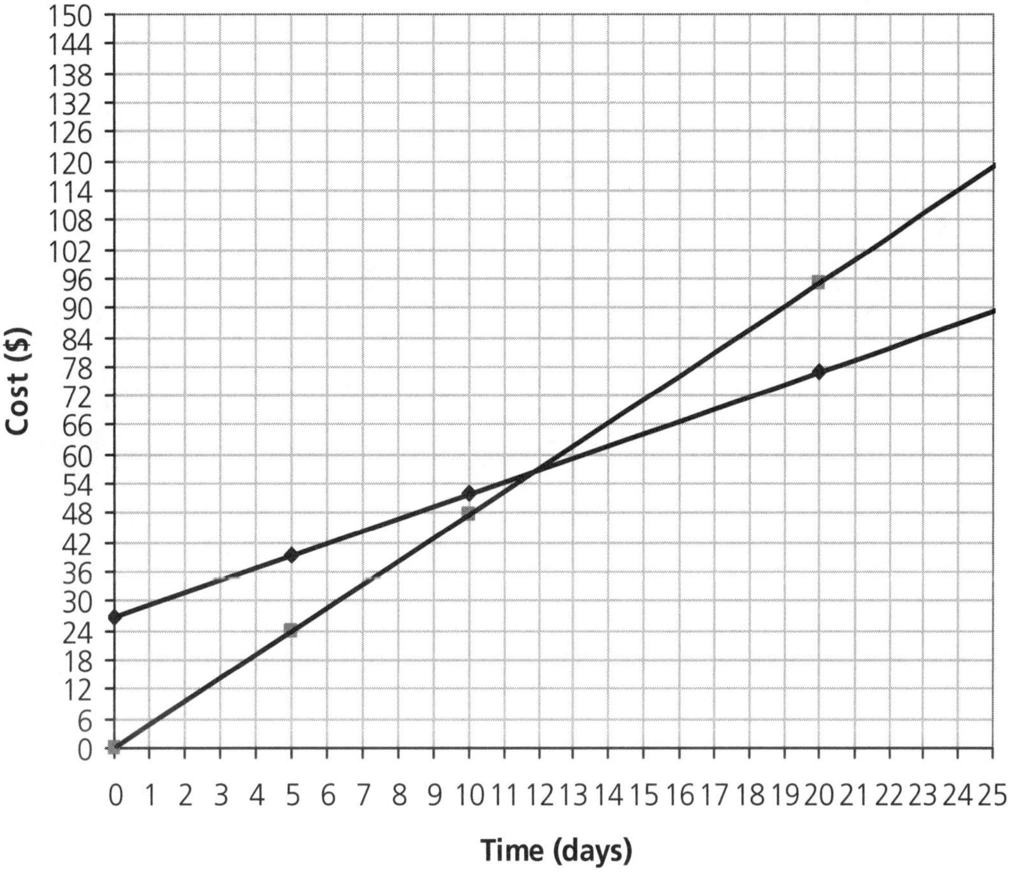

Task 4

Use the information you have gathered and analyzed in Tasks 1– 3 to write a paragraph explaining which company you will rent from and why. Write your answer in complete sentences.

If you are renting for less than 12 days you should rent from Computers For Students. If you are renting for more than 12 days, you should rent from Computronics. If you are renting for exactly 12 days the cost is equal.

Rectangle Area Quiz

Name: ______________________________

1. These rectangular boxes need to be wrapped. To find out how much wrapping paper you need, calculate the total surface area of each box. (Hint: Each box has six sides.)

a. 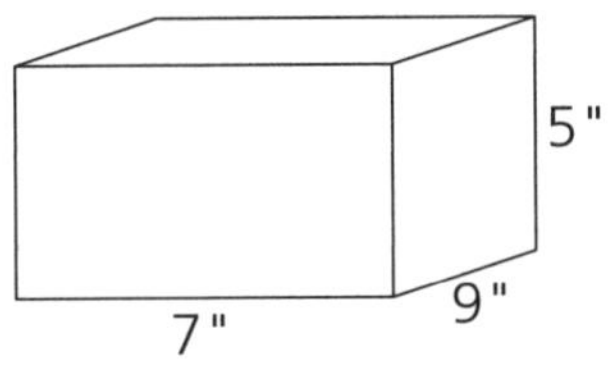

Top = 7(9)=63 u^2

Front = 7(5) = 35 u^2

Side = 9(5) = 45 u^2

SA = 2(63+35+45) = 286 u^2

b. 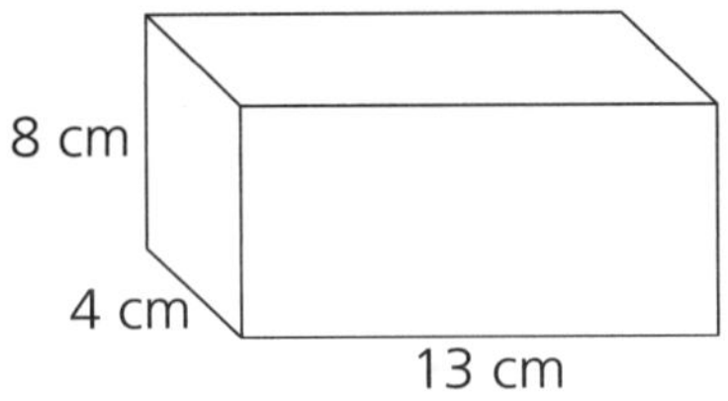

Top = 4(13)=52 u^2

Front = 8(13) = 104 u^2

Side = 8(4) = 32 u^2

SA = 2(52+104+32) = 376 u^2

c. 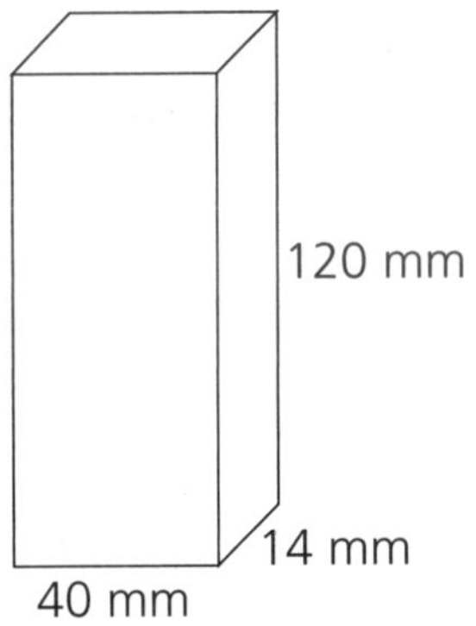

Top = 40(14)=560 u^2

Front = 40(120) = 4800 u^2

Side = 14(120) = 1680u^2

SA = 2(560+4800+1680) = 7040 u^2

2. In the figure, a side of the inner square measures 20 inches. A side of the outer square measures 30 inches. Find the area of the shaded region.

Outer square = 30(30) = 900 sq. ins.

Inner square = 20(20)= 400 sq. ins.

Shaded Region = 900 – 400 = 500 sq. ins.

3. In the figure, the outer rectangle measures 8 feet by 12 feet. The inner rectangle measures 5 feet by 7 feet. Find the area of the shaded region.

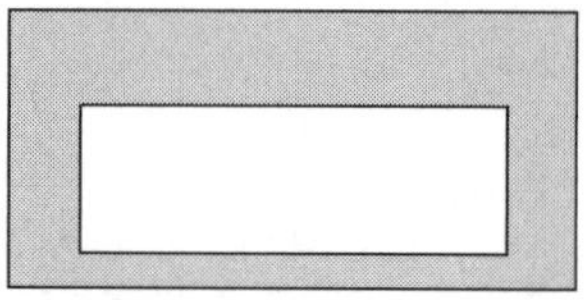

Outer = 8(12) = 96 sq. ft.; Inner = 5(7) = 35 sq. ft.;

Shaded region = 96 – 35 = 61 sq. ft.

4. Measure the figure with a centimeter ruler to find the area of the shaded region.

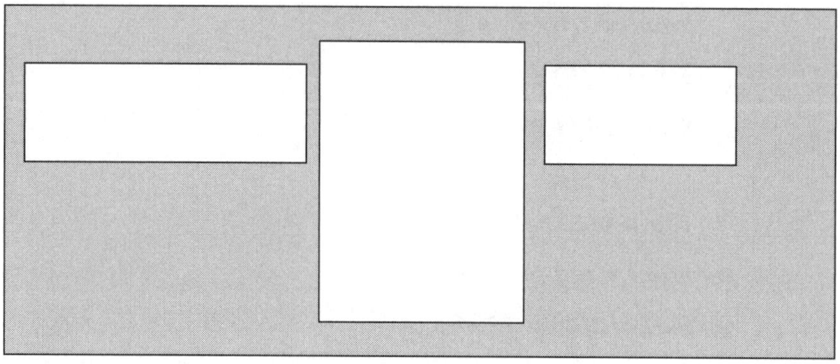

Outer Length = 10.9 cm &Width = 4.6 cm; Inner Rectangle Left Length = 3.7 cm & Width = 1.3 cm; Inner Rectangle Center Length = 2.7cm & Width = 3.7 cm; Inner Rectangle Right Length = 2.5 cm & Width = 1.3 cm.

Areas: Outer Rectangle = 50.14 cm^2; Inner Left Rectangle = 4.81 cm^2; Inner Center Rectangle = 9.99 cm^2; and Inner Right Rectangle = 3.25 cm^2.

Shaded Region = 50.24 – (3.25 + 9.99 + 4.81) = 32.09 cm^s.

5. The area of a rectangular field is 680 square miles. If the length of the field is 80 miles, find the width of the field. Write a sentence explaining how you arrived at your answer.

If you know the length of the field is 80 miles and the area is 680 square miles, then you know 80w = 680. Divide both sides by 80. The width is 8.5miles.

6. The area of a square garden is 529 square feet. Find the perimeter of the garden. Write a sentence explaining how you arrived at your answer.

   s^2 = 529 square feet

   s = 23

   Perimeter = 4(23) = 92 feet

   Take the square root to get the side length. Then multiply by 4.

7. You just bought a 50-square-foot roll of wrapping paper. The width of the roll is 1 foot. What are the dimensions of the largest rectangular box you can wrap with this roll of paper? Write a sentence explaining how you arrived at your answer.

   Answers will vary.

8. You are designing a red and black checkerboard banner, which will have an area of 576 square inches. All checkerboards have eight rows and eight columns of squares. How long is the rod you will need to hang the banner? How many square inches of red cloth will you need to make the banner? Write a sentence explaining how you arrived at your answer.

   s^2 = 576 square inches

   s = 24 inches

   The rod will need to be 24 inches. You will need 288 square inches of red cloth since half of the total area will be red.

Parallelogram Area Quiz

Name: ____________________________

1. Find the area of each parallelogram.

a. 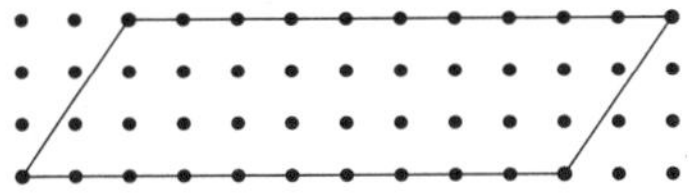

A = 10(3) = 30 u²

b. 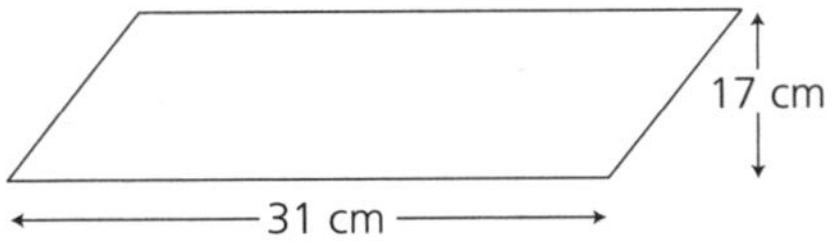

A = 17(31) = 527 u²

c. 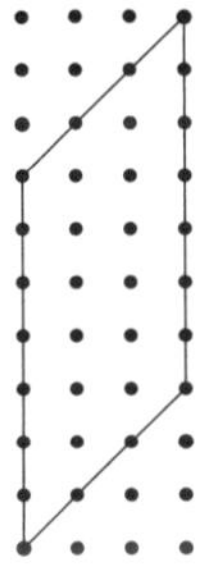

A = 7(3) = 21 u²

d. 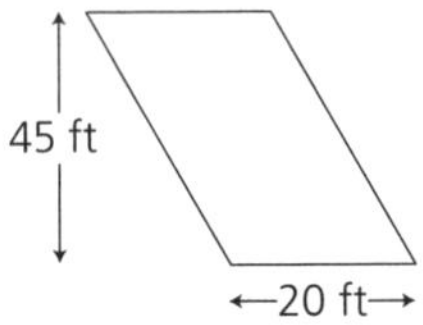

A = 20(45) = 900 u²

e. 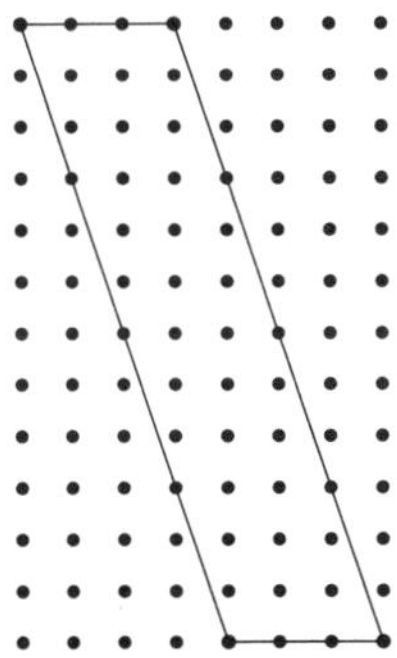

A = 3(12) = 36 u²

2. If the height of a parallelogram is 22 inches and the length of the base is 25 inches, find the area of the parallelogram. Show your work.

   A =22(25) = 550 in²

3. If the base and the height of a parallelogram are equal in measure, the parallelogram must be a square. Is this statement true or false? Explain your answer.

   False, see diagram provided.

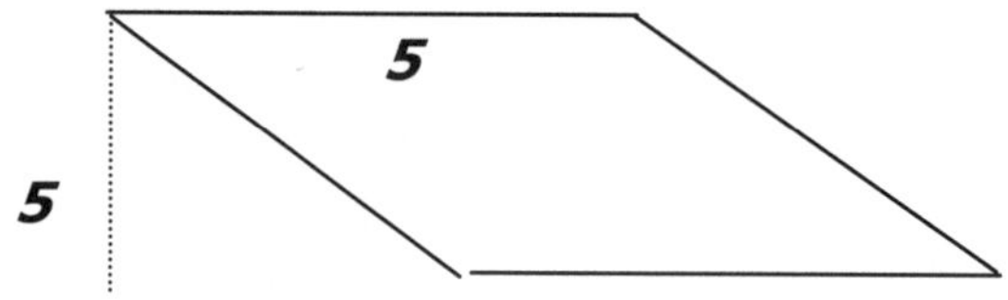

4. If the base and height of a parallelogram are doubled in size, the area will also double in size. Is this statement true or false? Explain your answer.

   $\boldsymbol{b_2 = 2b}$

   $\boldsymbol{h_2 = 2h}$

   $\boldsymbol{A_2 = (2b)(2h) = 4bh}$

   No, the area will quadruple.

5. The area of a parallelogram is 330 square feet and the base measures 22 feet. Find the height of the parallelogram. Show your work.

   22h = 330 square feet

   h = 15 feet

6. If the area of a parallelogram is 390 square yards and the height is 15 yards, find the length of the base of the parallelogram. Show your work.

   15 b = 390 square yards

   b = 26 yards

7. The area of a parallelogram is 200 square inches and the base is twice the measure of the height. Find the base and the height of the parallelogram. Show your work.

   h(2h) = 200 square inches

   $2h^2$ = 200 square inches

   h^2 = 100 square inches

   h = 10 inches

   Height is 10 inches, base is 20 inches.

8. If the area of a parallelogram is 192 square inches and the base is three times the measure of the height, find the base and the height. Show your work.

   h(3h) = 192 square inches

   $3h^2$ = 192 square inches

   h^2 = 64 square inches

   h = 8 inches

   Height is 8 inches, base is 24 inches.

Triangle Area Quiz

Name: ____________________________

1. Find the area of both triangles.

a.

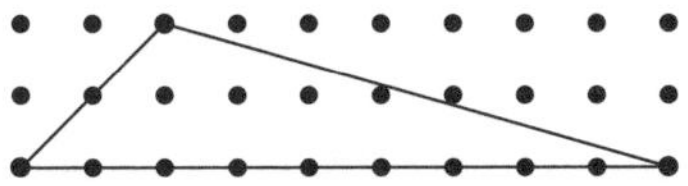

$A = \frac{1}{2}(2)(9) = 9$ square units

b.

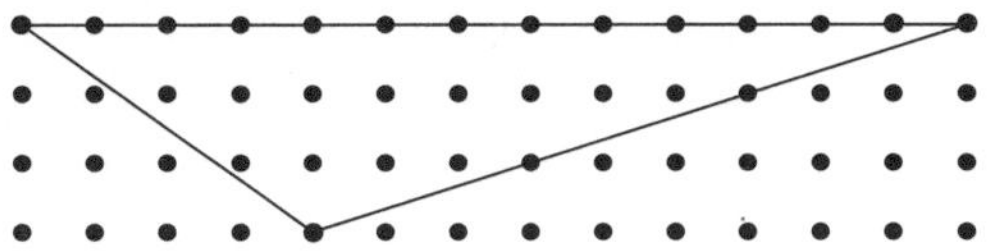

$A = \frac{1}{2}(13)(3) = 19.5\ u^2$

2. Find the area of each figure.

a.

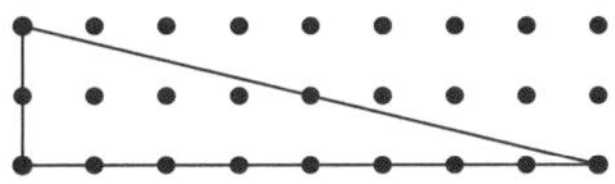

$A = \frac{1}{2}(8)(2) = 8$ sq. u.

b.

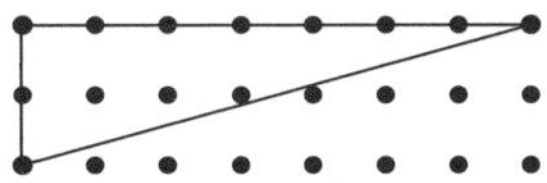

$A = \frac{1}{2}(7)(2) = 7\ u^2$

c.

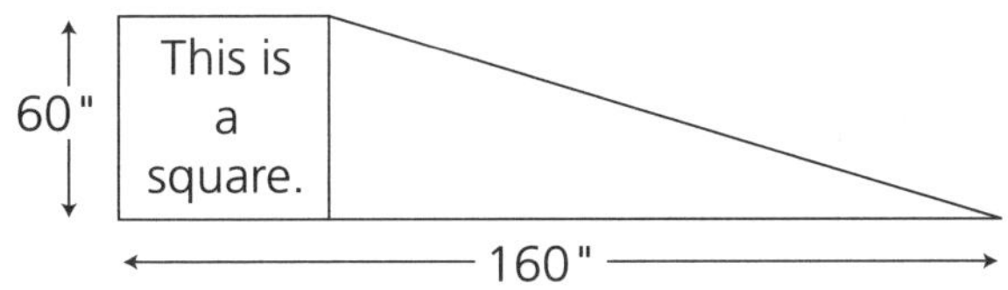

$A = (60)(60) + (\frac{1}{2})(60)(100)$

$A = 3600 + 3000 = 3600\ in^2$

3. You are making a nylon kite. The kite sticks are perpendicular to each other, and the longer stick intersects the middle of the shorter stick. If the height of the kite is 42 inches and its greatest width is 30 inches, find the amount of nylon you will need to make the kite. (Hint: find the amount of nylon you will need to make the kite. (Hint: Find the area.)

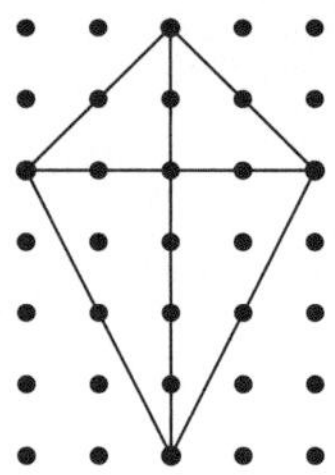

A = ½ (15)(42) + ½ (15)(42)

A =(½)(42)(30) = 630 in²

4. You are designing a triangular pennant for your school. If you have only 330 square inches of fabric and you want the pennant, when suspended from its base, to hang down 44 inches, what will be the length of the base?

½ b(44) = 330 square inches

22b = 330 square inches

b = 15 inches

5. The perimeter of a baseball diamond is 360 feet. (A baseball diamond is actually a square.) How many square feet of grass must be mowed inside the triangle formed by the pitcher's mound, second base, and third base? Write a sentence explaining how you arrived at your answer.

Area of the square = 90(90) = 8100 ft²

Area of the Triangle = ¼(8100) = 2025 ft²

Sample Response:

I found the area of the baseball diamond based on knowing the perimeter and that it is actually a square. I divided the whole area by four to find the area of the triangle where the grass has to be mowed.

6. The sail on a boat is triangular and required 540 square feet of nylon to make. If the base of the sail is 36 feet, find its height.

   ½(36)h = 540 square feet

   18h = 540 square feet

   h = 30 feet

   The height of the sail is 30 feet.

7. A triangular window in a beach house is 48 inches in height and contains 720 square inches of glass. Find the measure of the base of the window.

   720 = ½ b(48)

   720 = 24b

   30 = b

   The base of the window is 30 inches.

Trapezoid Area Quiz

Name: ____________________________

1. Find the area of each trapezoid.

a.

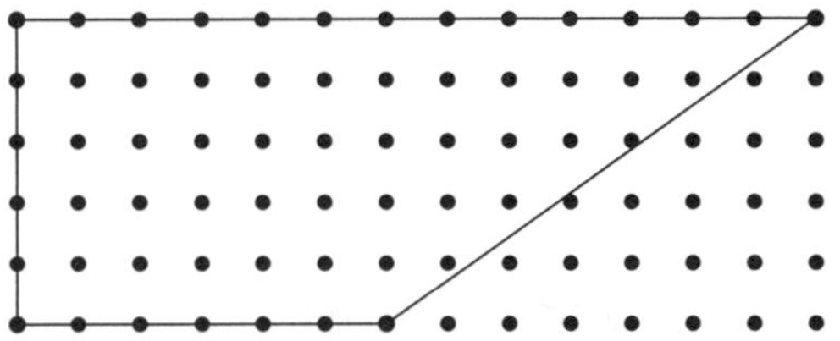

$$A = \frac{1}{2}h(b_1 + b_2)$$

$$A = \frac{1}{2} \bullet 5 \bullet (6 + 13)$$

$$A = \frac{1}{2} \bullet 5 \bullet 19 = 47.5 \text{ units}^2$$

b.

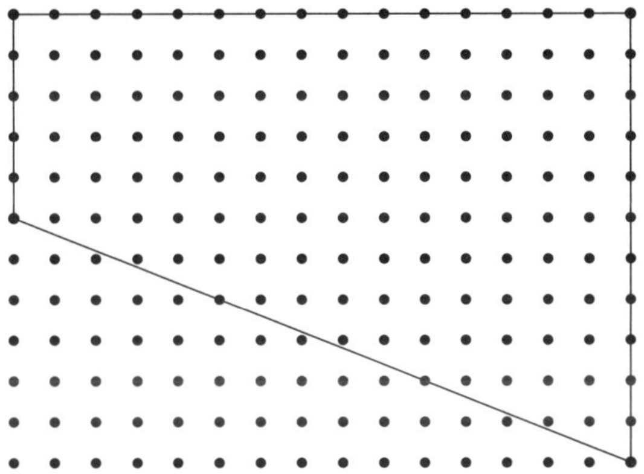

$$A = \frac{1}{2}h(b_1 + b_2)$$

$$A = \frac{1}{2} \bullet 15 \bullet (5 + 11)$$

$$A = \frac{1}{2} \bullet 15 \bullet 16$$

$$A = 120 \text{ units}^2$$

c.

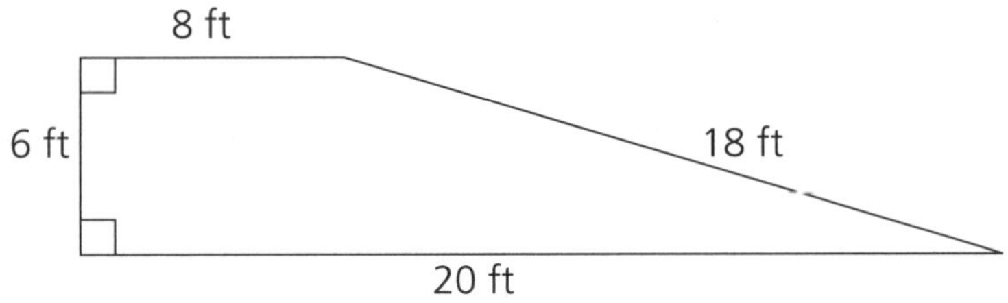

$$A = \frac{1}{2}h(b_1 + b_2)$$

$$\frac{1}{2} \bullet 6 \bullet (8 + 20)$$

$$\frac{1}{2} \bullet 6 \bullet 28 = 84 ft^2$$

d.

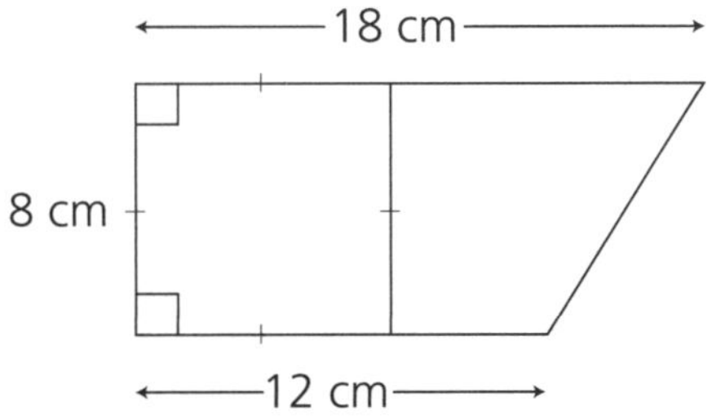

$$A = \frac{1}{2}h(b_1 + b_2)$$

$$A = \frac{1}{2} \bullet 8 \bullet (18 + 12)$$

$$A = \frac{1}{2} \bullet 8 \bullet 30 = 120 cm^2$$

2. If the height of a trapezoid is 5 inches, and the bases measure 6 inches and 11 inches, find the area of the trapezoid. Show your work.

   $A = \frac{1}{2}h(b_1 + b_2)$

   $A = \frac{1}{2} \bullet 5 \bullet (6 + 11)$

   $A = 42.5 \text{ inches}^2$

3. Can the bases of a trapezoid ever be the same length? Explain why or why not.

   No. If the bases of a trapezoid were the same length, it would be a parallelogram.

4. If the bases of a trapezoid measure 20 feet and 42 feet, and the height is 24 feet, find the area of the trapezoid.

   $A = \frac{1}{2}h(b_1 + b_2)$

   $A = \frac{1}{2} \bullet 24 \bullet (20 + 42)$

   $A = 744 \text{ ft}^2$

5. The area of a trapezoid is 396 square yards, and the bases measure 13 yards and 20 yards. Find the height of the trapezoid. Show your work.

   $A = \frac{1}{2}h(b_1 + b_2)$

   $396 = \frac{1}{2} \bullet h \bullet (13 + 20)$

   $396 = \frac{1}{2} \bullet h \bullet 33$

   $396 = 16.5h$

   $h = \frac{396}{16.5} = 24 \text{ yd}$

6. The area of a trapezoid is 323 square miles, one base measures 26 miles, and the height is 24 miles. Find the length of the second base of the trapezoid. Show your work.

$$A = \frac{1}{2}h(b_1 + b_2)$$

$$323 = \frac{1}{2} \bullet 24 \bullet (26 + b_1)$$

$$323 = 12 \bullet (26 + b_1)$$

$$26.9 \approx 26 + b_1$$

$$b_1 \approx 26.9 - 26 = 0.9 \text{ miles}$$

7. The area of a trapezoid is 150 square meters. The height is 20 meters, and one base is three meters longer than the other base. Find the length of both bases. Show your work.

$$A = \frac{1}{2}h(b_1 + b_2) \qquad b_1 = b_2 + 3$$

$$150 = \frac{1}{2} \bullet 20 \bullet (b_2 + 3 + b_2)$$

$$150 = 10 \bullet (2b_2 + 3)$$

$$15 = 2b_2 + 3$$

$$12 = 2b_2$$

$$b_2 = 6 \text{ meters} \quad \text{and} \quad b_1 = 6 + 3 = 9 \text{ meters}$$

8. The area of a trapezoid is 504 square feet and the height is 12 feet. Find the length of both bases if the first base is twice the length of the second base. Show your work.

$$A = \frac{1}{2}h(b_1 + b_2) \qquad b_1 = 2b_2$$

$$504 = \frac{1}{2} \bullet 12 \bullet (2b_2 + b_2)$$

$$504 = 6 \bullet 3b_2$$

$$84 = 3b_2$$

$$b_2 = 28 \text{ ft} \quad \text{and} \quad b_1 = 2 \bullet 28 = 56 \text{ ft}$$

Polygon Area Quiz

Name: ______________________

1. Find the area of each regular polygon. ***In the formula for area, p is the perimeter of the polygon and a is the length of the apothem.***

a.

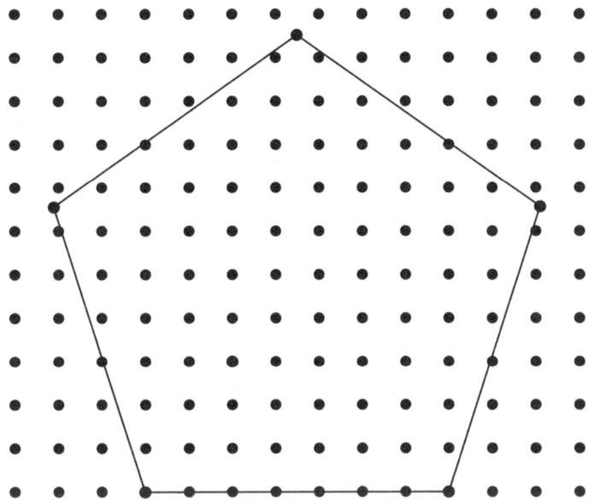

$A = \frac{1}{2}pa$

$A = \frac{1}{2} \bullet (7 \bullet 5) \bullet 5$

$A = \frac{1}{2} \bullet 35 \bullet 5$

$A = 87.5 \text{ units}^2$

b.

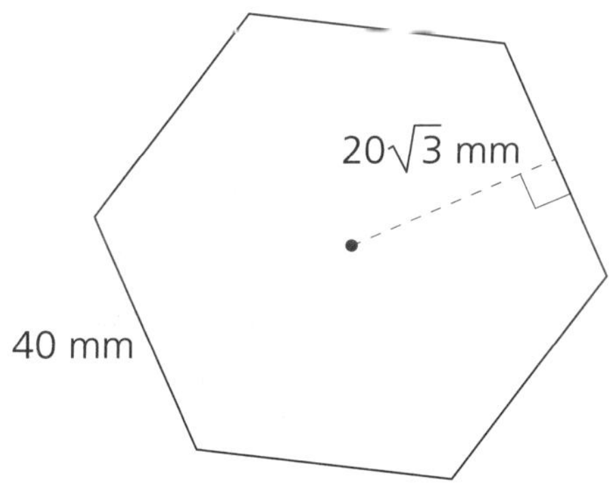

$A = \frac{1}{2}pa$

$A = \frac{1}{2} \bullet (40 \bullet 6) \bullet 20\sqrt{3}$

$A = \frac{1}{2} \bullet 240 \bullet 20\sqrt{3}$

$A \approx 4156.92 \text{ mm}^2$

c.

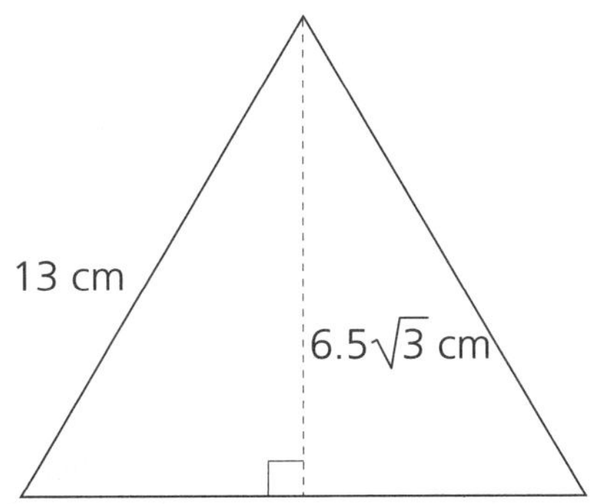

$A = \frac{1}{2}bh$

$A = \frac{1}{2} \bullet 13 \bullet 6.5\sqrt{3}$

$A = 6.5 \bullet 6.5\sqrt{3}$

$A \approx 73.18 \text{ cm}^2$

2. If the distance from the center to a side of a regular heptagon is 20.76 inches and each side measures 20 inches, find the area of the heptagon. Show your work.

$$A = \frac{1}{2}pa$$

$$A = \frac{1}{2} \bullet (20 \bullet 7) \bullet 20.76$$

$$A = \frac{1}{2} \bullet 140 \bullet 20.76$$

$$A = 1453.2 \text{ inches}^2$$

3. If the distance from the center to a side of a stop sign is 36.21 inches and the length of each side is 30 inches, find the area of the stop sign. Show your work.

$$A = \frac{1}{2}pa$$

$$A = \frac{1}{2} \bullet (30 \bullet 8) \bullet 36.21$$

$$A = \frac{1}{2} \bullet 240 \bullet 36.21$$

$$A = 4345.2 \text{ inches}^2$$

4. If the apothem (distance from the center to a side) of a regular decagon is 33.85 meters and each side measures 22 meters, find the area of the decagon. Show your work.

$$A = \frac{1}{2}pa$$

$$A = \frac{1}{2} \bullet (22 \bullet 10) \bullet 33.85$$

$$A = \frac{1}{2} \bullet 220 \bullet 33.85$$

$$A = 3723.5 \text{ m}^2$$

5. If the area of a regular nonagon is 1211.49 square yards and the apothem is 19.23 yards, find the length of each side. Show your work.

$A = \frac{1}{2}pa$

$1211.49 = \frac{1}{2} \bullet [(\text{length of each side}) \bullet 9] \bullet 19.23$

$1211.49 = 86.535 \bullet (\text{length of each side})$

14 yd = length of each side

6. A quilt is made from interlocking regular hexagons. The distance from the center to a side of each hexagon is 1.7 inches and the perimeter of each hexagon is 12 inches. How many regular hexagons would be needed to make a quilt measuring 8 feet by 8 feet? Show your work.

Area of each hexagon:

$A = \frac{1}{2}pa$

$A = \frac{1}{2} \bullet 12 \bullet 1.7$

$A = 10.2\ in^2$

Area to be covered:

$A = bh$

$A = 8 \bullet 8$

$A = 64\ ft^2$

$A = 64\ ft^2 \bullet 144 \frac{in^2}{ft^2} = 9216\ in^2$

Number of hexagons to cover the bed:

$9216\ in^2 \div 10.2\ in^2 \approx 903.5$, ***or 904 hexagons***

An alternate solution uses the length and width of the quilt instead of the area of the hexagonal pieces. The hexagons tessellate the quilt. Two hexagonal pieces joined use 6" in length, so 4 pieces will measure 1 foot. This means 32 pieces are needed to make the length 8 feet. Also, 32 pieces are needed to make the width 8 feet. 32*32 = 1024 hexagonal pieces.

7. If the side of a square is 20 meters long, what is:

   a. The area of the square?

      A = bh

      $A = 20 \bullet 20 = 400\ m^2$

   b. The length of the distance from the center to a side (apothem)?

      $A = \frac{1}{2}pa$

      $400 = \frac{1}{2} \bullet (20 \bullet 4) \bullet a$

      $400 = 40a$

      $a = 10\ m$

      Also, in a square, the distance from center to a side is ½ the length of the side, so the apothem must be 10 m.

   c. If the length of a side doubles, how does that affect the area of the square? (Does it double?) How does it affect the length of the apothem?

If length of a side is doubled:	***$s = 2 \bullet 20 = 40\ m$***
then the area is quadrupled:	***$A = s^2$***
	$A = 40^2$
	$A = 1600\ m^2 = \underline{4} \bullet 400\ m^2$
and the apothem is doubled:	***$A = \frac{1}{2}pa$***
	$1600 = \frac{1}{2} \bullet (40 \bullet 4) \bullet a$
	$a = 20\ m = \underline{2} \bullet 10\ m$

The apothem is still half of the length of the base, so it must double if the base doubles.

Circle Area Quiz

Name: ____________________________

1. Find the area of each figure. (What appears to be a semicircle is just that.)

a.

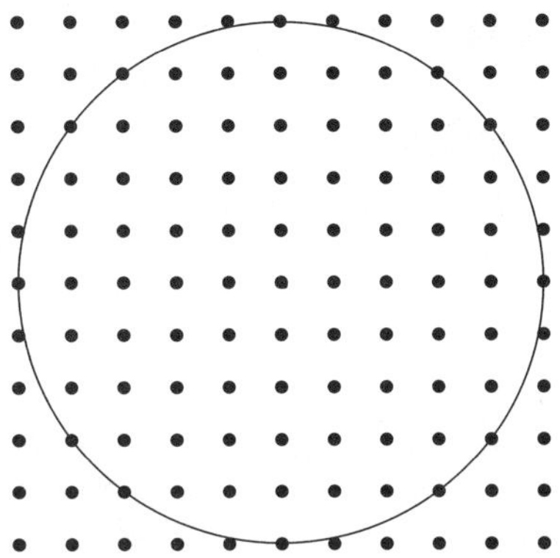

$\boldsymbol{A = \pi * 5^2}$

$\boldsymbol{A \approx 78.5 \text{ units}^2}$

b.

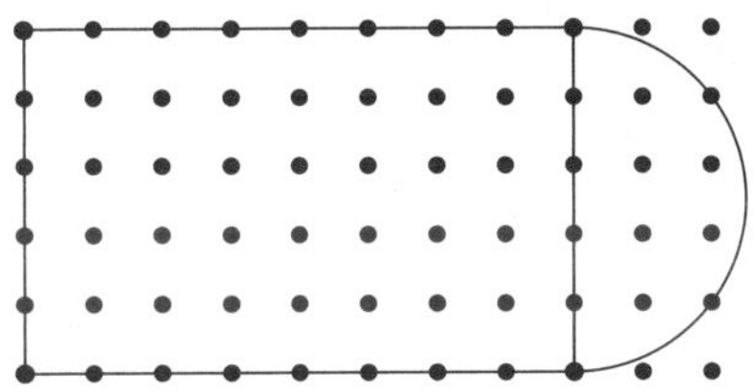

$\boldsymbol{A = \frac{1}{2}\pi\,(2.5)^2 + 5*8}$

$\boldsymbol{A \approx 9.81 + 40}$

$\boldsymbol{A \approx 49.81 \text{ units}^2}$

c.

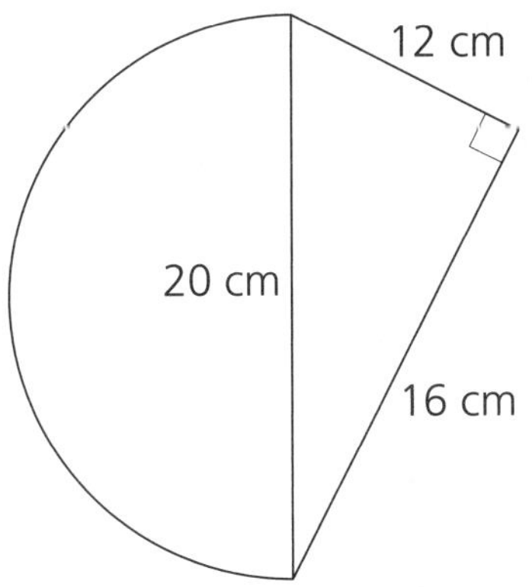

$\boldsymbol{A = \frac{1}{2}\pi\,(10)^2 + \frac{1}{2}(12)(16)}$

$\boldsymbol{A \approx 157 + 96}$

$\boldsymbol{A \approx 253 \text{ cm}^2}$

2. A spotlight rotates 360° and emits a light beam that is visible up to 1500 feet from the light. Find the area covered by the light. Show your work.

   $A = \pi r^2$

   $A \approx 3.14 \bullet 1500 \bullet 1500$

   $A \approx 7{,}065{,}000 \text{ feet}^2$

3. A lawn sprinkler spins 360° and has a 4-foot spray. Find the area of lawn the sprinkler will water.

   $A = \pi r^2$

   $A \approx 3.14 \bullet 4 \bullet 4$

   $A \approx 50.24 \text{ feet}^2$

4. A circular bed of flowers has a circumference of 8 feet. Find the area of the flower bed.

   $C = 2\pi r$ $\quad A = \pi r^2$

   $8 = 2\pi r$ $\quad A \approx 3.14 \bullet 1.27 \bullet 1.27$

   $r \approx 1.27 \text{ ft}$ $\quad A \approx 5.06 \text{ ft}^2$

5. In the figure, the area of the square is 36 square feet. The diagonal of the square is $\sqrt{72}$ feet. Find the area of the shaded region.

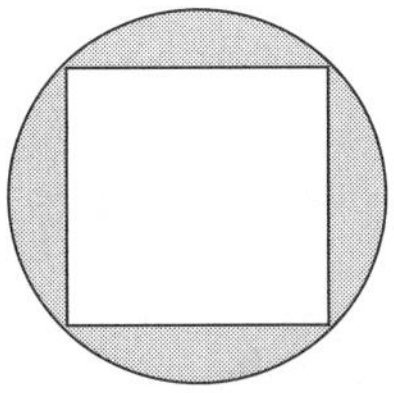

Since the area of the square is 36 ft^2, the length of each side is $\sqrt{36}$, or 6 ft.
The diagonal of the square is also the diameter of the circle: $\sqrt{72}$, or about 8.49 ft.
The radius of the circle is half the diameter: $\frac{8.49}{2}$, or about 4.24 ft.
The area of the circle is about $\pi \bullet 4.24^2$, or 56.45 ft^2.
The area of the shaded region is the difference between the area of the circle and the area of the square: 56.45 ft^2 – 36 ft^2 = 20.45 ft^2 (approximately)

6. If it takes 28.26 square inches of plastic to manufacture the circular lid for a paper cup, what is the radius of the cup? Show your work.

   $A = \pi r^2$

   $28.26 \approx 3.14 \bullet r^2$

   $9 \approx r^2$

   $r \approx 3$ inches

7. In a circular baby pool, 3215.36 square feet of water is exposed to the air. To make a cover for the pool you need to know its diameter. Find the diameter of the pool. Show your work.

   $A = \pi r^2$

   $3215.36 \approx 3.14 \bullet r^2$

   $1024 \approx r^2$

   $r \approx 32$ inches

   The diameter is approximately 64 inches.

8. In this figure, the area of the square is 121 square feet. The circle is just touching each side of the square. Find the area of the shaded region. Show your work.

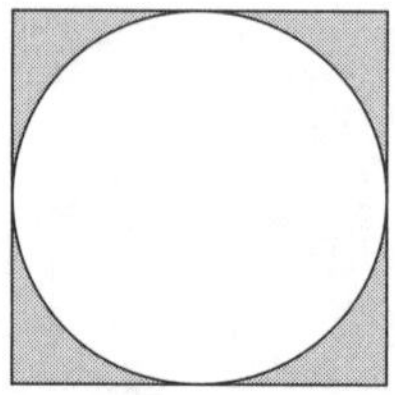

Since the area of the square is 121 ft^2, the length of each side is $\sqrt{121}$, or 11 ft, which is also the diameter of the circle.

The radius of the circle is half the diameter: $\frac{11}{2}$, or 5.5 ft.

The area of the circle is about $\pi \bullet 5.5^2$, or about 94.98 ft^2.

The area of the shaded region is the difference between the area of the square and the area of the circle: 121 ft^2 – 94.98 ft^2 = 26.02 ft^2 (approximately)

9. This figure is composed of a semicircle and a square. The area of the square is 169 square feet. Find the perimeter of the figure.

The perimeter is the sum of the lengths of three sides of the square plus the circumference of the semicircle.

Since the area of the square is 169 ft^2, the length of each side is $\sqrt{169}$, or 13 ft, which is also the diameter of the semicircle.

Circumference of the semicircle: $\frac{1}{2}\pi d \approx \frac{1}{2} \bullet 3.14 \bullet 13 = 20.41$ ft

Perimeter of the figure: $13 \bullet 3 + 20.41 = 59.41$ ft (approximately)

Pythagorean Theorem Test

Name: ______________________________

For problems 1–4:

- Draw a picture
- Show your work
- Write a complete sentence stating your solution

1. If a small rectangular box measures 10 cm by 5 cm by 2 cm, what is the longest toothpick that will fit in the box without bending?

Length of the diagonal across the bottom face:

$\sqrt{a^2 + b^2} = \sqrt{100 + 25} \approx 11.18$ ***cm***

Length of the diagonal of the box:

$\sqrt{125 + 4} \approx 11.36$ ***cm***

The longest toothpick that will fit in the box is about 11.36 centimeters.

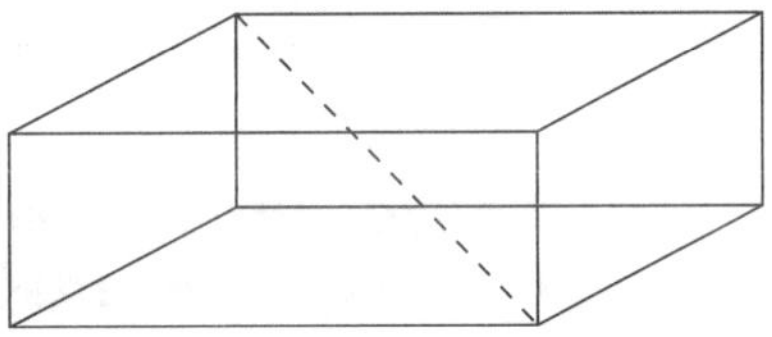

(Students could also use the 3-D equivalent of the Pythagorean Theorem, $\sqrt{a^2 + b^2 + c^2}$***, to solve this problem.)***

2. A box of long-stemmed roses just arrived at your house. If the rectangular box measures 3 feet by 1 foot by 0.5 feet, find the maximum possible length of one rose. (Assume that the rose and the stem of the rose are not bent.)

Length of the diagonal across the bottom face:

$\sqrt{a^2 + b^2} = \sqrt{9 + 1} \approx 3.16$ ***feet***

Length of the diagonal of the box:

$\sqrt{10 + 0.25} = \sqrt{10.25} \approx 3.2$ ***ft***

The longest rose that will fit in the box is about 3.2 feet.

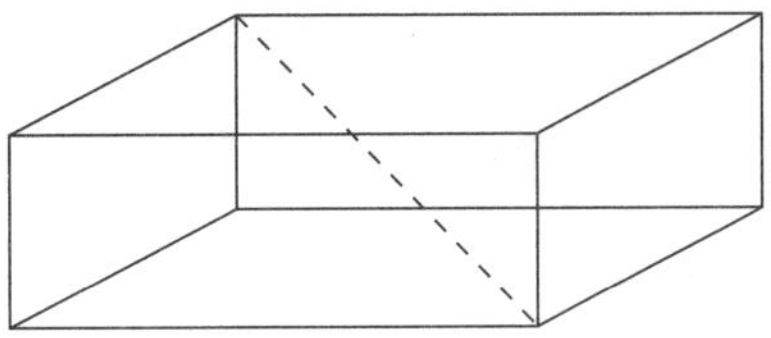

3. A square sheet of glass measures 8 feet on each side. Describe what will happen if you try to get it through a rectangular door frame that measures 4 feet by 7 feet. Mathematically prove whether the glass *will* or *will not* fit through the door. (Do not concern yourself with the thickness of the glass.)

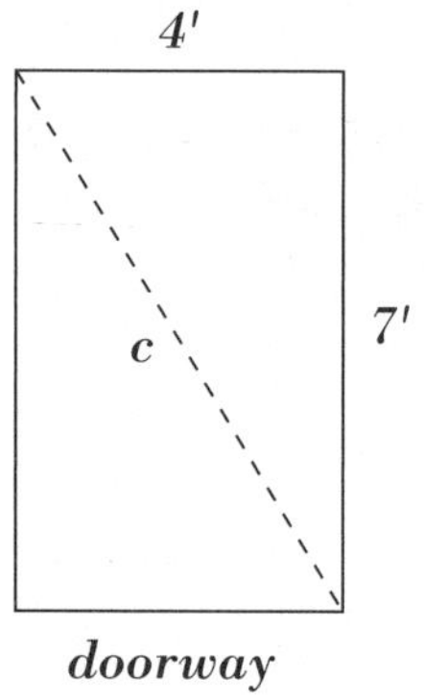

$4^2 + 7^2 = c^2$

$16 + 49 = c^2$

$65 = c^2$

$c \approx 8.06$ ft

The 8-foot square glass will fit through the door frame diagonally, with 0.06 feet to spare.

4. A letter measuring 33 cm by 4.9 cm was in your mailbox today. If your mailbox measures 30 cm by 15 cm by 5 cm, did the letter fit entirely in the mailbox without folding or bending?

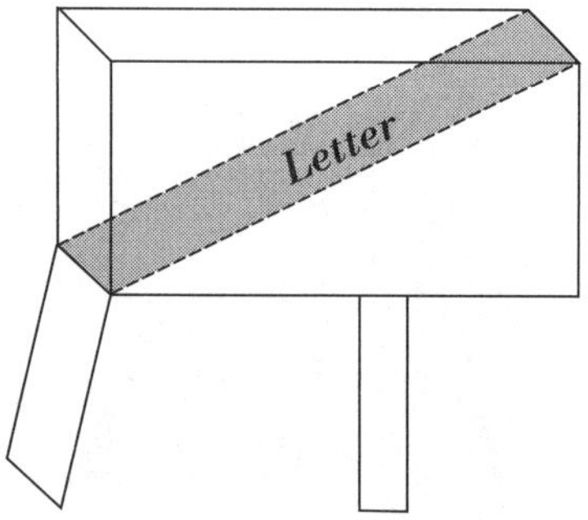

Unlike in the other problems, you do not want to find the 3-dimensional diagonal because a letter is not shaped like a stick or line. Instead, since you know that the letter will only fit if its width goes across the 5 cm dimension of the box, you will have to find out what the length of the diagonals of the mailbox are as in the picture.

The length of the diagonals on the sides of the box are:

$\sqrt{30^2 + 15^2} = \sqrt{900 + 225} = \sqrt{1125} \approx 33.54$ cm

The longest letter that will fit in the mailbox is a little over 33 and a half centimeters, so this letter will fit.

For problems 5–10, each triangle is a right triangle. Use the Pythagorean Theorem to find the length of the unknown side. Show all or your work in solving each problem.

5.

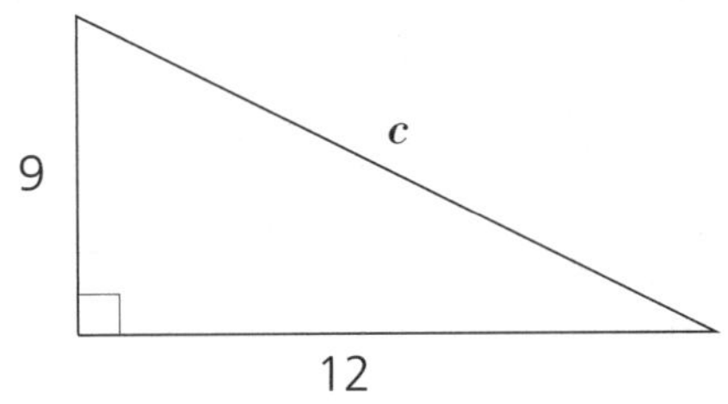

$a^2 + b^2 = c^2$

$9^2 + 12^2 = c^2$

$81 + 144 = c^2$

$225 = c^2$

$c = 15$

6.

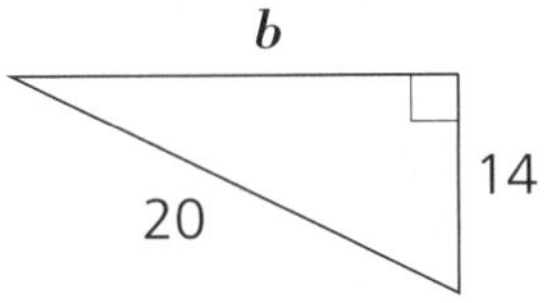

$a^2 + b^2 = c^2$

$14^2 + b^2 = 20^2$

$196 + b^2 = 400$

$b^2 = 204$

$b \approx 14.28$

7.

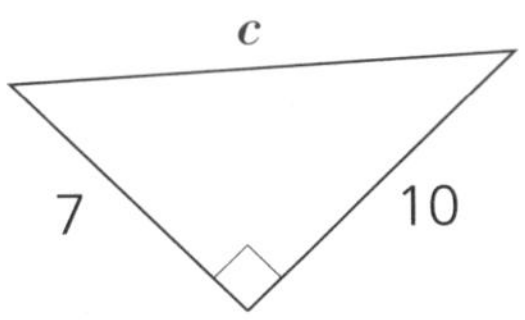

$a^2 + b^2 = c^2$

$7^2 + 10^2 = c^2$

$49 + 100 = c^2$

$149 = c^2$

$c \approx 12.21$

8. 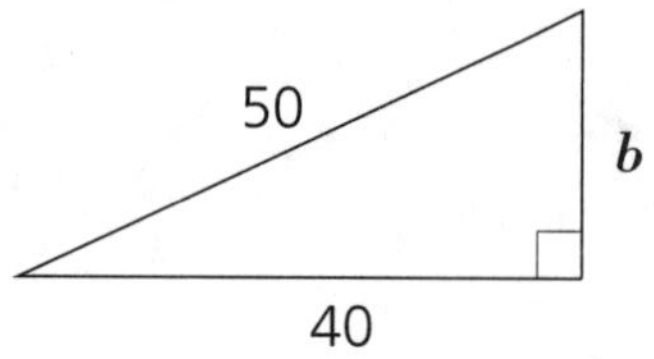

$$a^2 + b^2 = c^2$$
$$40^2 + b^2 = 50^2$$
$$1{,}600 + b^2 = 2{,}500$$
$$b^2 = 900$$
$$b = 30$$

9. 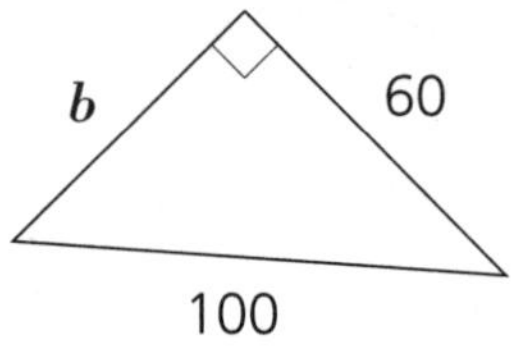

$$a^2 + b^2 = c^2$$
$$60^2 + b^2 = 100^2$$
$$3{,}600 + b^2 = 10{,}000$$
$$b^2 = 6{,}400$$
$$b = 80$$

10. 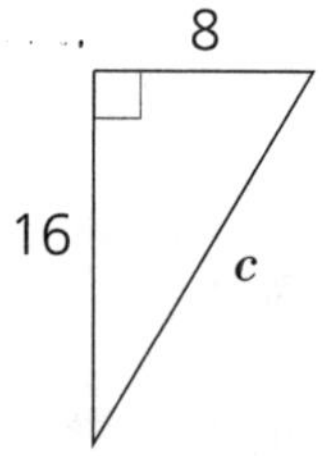

$$a^2 + b^2 = c^2$$
$$8^2 + 16^2 = c^2$$
$$64 + 256 = c^2$$
$$320 = c^2$$
$$c \approx 17.89$$

Special Right Triangle Quiz 1 (Form A)

Name: ____________________________

Show the work required to solve each problem.

1. If the legs of this isosceles triangle each measure 24 inches, find the length of the hypotenuse.

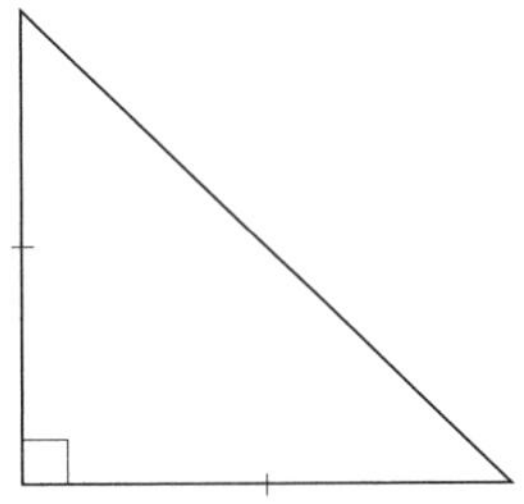

hypotenuse = leg $\bullet\sqrt{2}$

hypotenuse $= 24 \bullet \sqrt{2} \approx 33.94$ ***inches***

2. Find *a*.

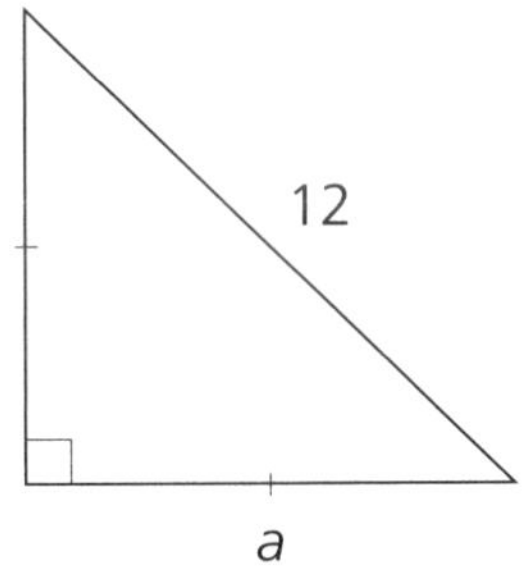

hypotenuse $= a \bullet \sqrt{2}$

$12 = a \bullet \sqrt{2}$

$a = \frac{12}{\sqrt{2}} \approx 8.49$

3. The area of this square is 36 cm^2. Find the length of its diagonal.

Since the area of the square is 36 cm^2***, the length of each side is*** $\sqrt{36}$***, or 6 cm.***

hypotenuse = leg $\bullet\sqrt{2}$

hypotenuse $= 6 \bullet \sqrt{2} \approx 8.49$ ***inches***

Special Right Triangle Quiz 1 (Form B)

Name: ____________________________

Show the work required to solve each problem.

1. If the hypotenuse of this isosceles triangle measures 24 inches, find the length of the each leg.

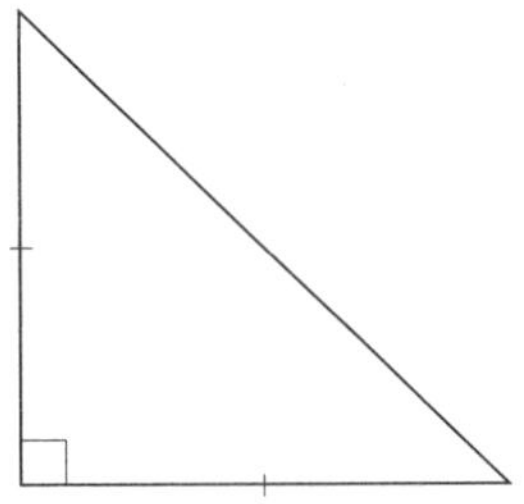

hypotenuse $= leg \cdot \sqrt{2}$

24 $= leg \cdot \sqrt{2}$

leg $= \frac{24}{\sqrt{2}}$

leg ≈ 16.97

2. Find *a*.

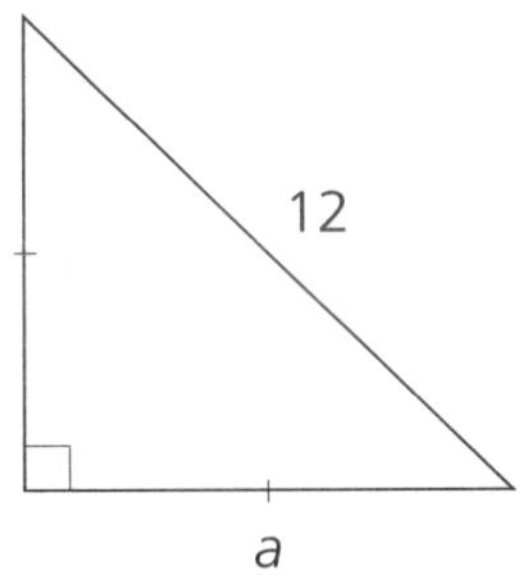

hypotenuse $= a \cdot \sqrt{2}$

12 $= a \cdot \sqrt{2}$

a $= \frac{12}{\sqrt{2}}$

a ≈ 8.49

3. If the diagonal of this square measures 20 cm, find its area.

20 $= leg \cdot \sqrt{2}$

leg $= \frac{20}{\sqrt{2}}$

leg ≈ 14.14 ***cm***

Each side of the square is about 14.14 cm.

Area of the square:

$14.14 \cdot 14.14 \approx 199.94\ cm^2$

Special Right Triangle Quiz 1 (Form C)

Name: ____________________________

Show the work required to solve each problem.

1. Find the area of this figure using the information given on the diagram. (The figure is composed of a right triangle and a semicircle.)

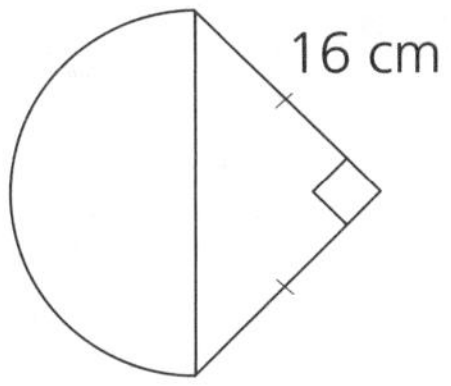

$a^2 + b^2 = c^2$

$16^2 + 16^2 = c^2$

$256 + 256 = c^2$

$512 = c^2$

$c \approx 22.63$ cm

The diameter of the semicircle is about 22.63 cm, so its radius is about 11.32 cm.

Area of the semicircle:

$A = \frac{1}{2}\pi r^2$

$A \approx 0.5 \bullet 3.14 \bullet 11.32 \bullet 11.32$

$A \approx 201.19\ cm^2$

Area of the triangle:

$A = \frac{1}{2}bh$

$A = \frac{1}{2} \bullet 16 \bullet 16$

$A = 128\ cm^2$

Area of the figure $\approx 201.19 + 128 = 329.19\ cm^2$

2. If the area of the square is 121 square inches, find the perimeter of this figure. (The figure is composed of a square and a semicircle.)

Since the area of the square is 121 square inches, the length of each side of the square is $\sqrt{121}$, or 11 inches.
The diameter of the semicircle is a side of the square, so it is also 11 inches long.
Circumference of the semicircle:

$C = \frac{1}{2}(2\pi r) = \pi r \approx 3.14 \bullet 5.5 = 17.27$ in

Perimeter of the entire figure:

$P = 3 \bullet 11 + 17.27 = 50.27$ in (approximately)

Special Right Triangle Quiz 2 (Form A)

Name: ______________________

Show the work required to solve each problem.

1. If the measure of the hypotenuse in this 30-60-90 triangle is 18 meters, find the length of sides *a* and *b*.

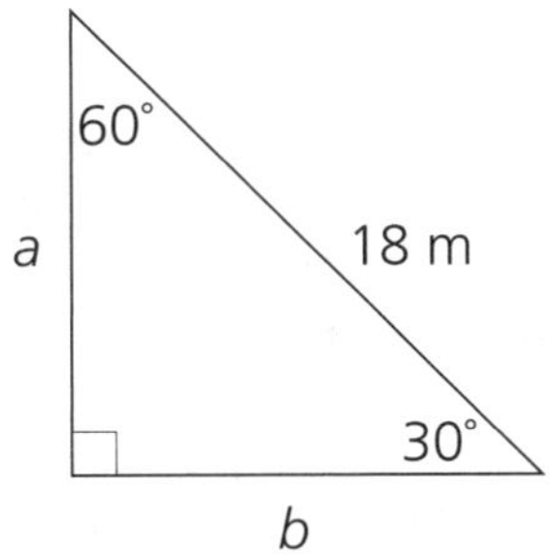

The length of the side opposite the 30° angle is half the length of the hypotenuse:

$$a = \frac{1}{2} \bullet 18 = 9 \text{ m}$$

The length of the side opposite the 60° angle is the length of the short leg times $\sqrt{3}$:

$$b = 9\sqrt{3} \approx 15.59 \text{ m}$$

2. In the triangle below, if the measure of the side opposite the 30° angle is 15 feet, find the length of sides *b* and *c*.

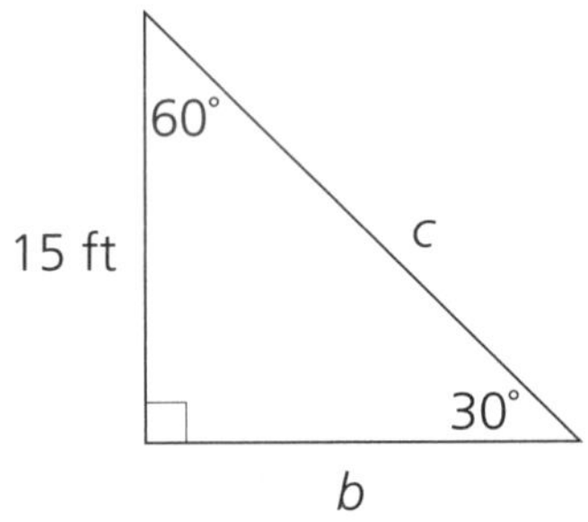

The length of the hypotenuse is double the length of the side opposite the 30° angle:

$$2 \bullet 15 = c$$

$$c = 30 \text{ ft}$$

The length of the side opposite the 60° angle is the length of the short leg times $\sqrt{3}$:

$$b = 15\sqrt{3} \approx 25.98 \text{ ft}$$

3. In the triangle below, if the measure of the side opposite the 60° angle is 9 millimeters, find the length of sides *a* and *c*.

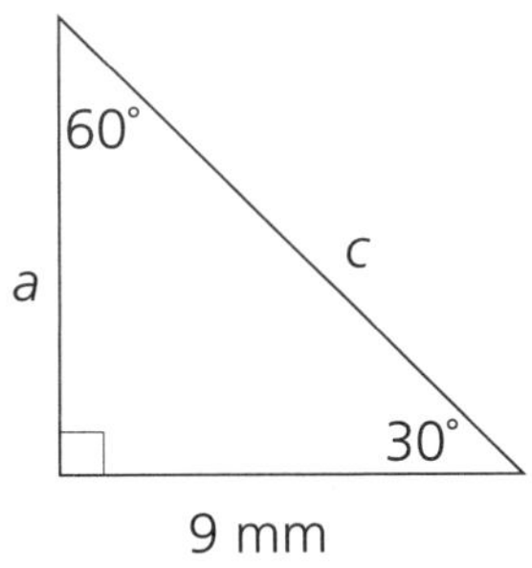

The length of the side opposite the 30° angle is the length of the side opposite the 60° angle divided by the $\sqrt{3}$:

$$a = 9/\sqrt{3} = 5.195 \text{ mm (approximately)}$$

The length of the hypotenuse is double the length of the short leg:

$$2(9/\sqrt{3}) \approx 10.39 \text{ mm}$$

4. A microwave dish is situated on top of a tower, and the signal travels from the dish to your house so you can watch cable TV. The angle of elevation from your house to the tower measures 30° and the distance from your house to the tower is 22 miles. Find the height of the tower and the distance the signal travels.

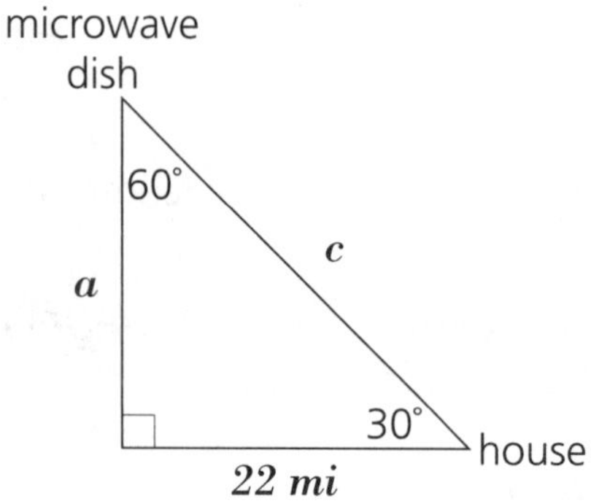

The length of the side opposite the 30° angle is the length of the side opposite the 60° angle divided by the $\sqrt{3}$:

$$a = 22/\sqrt{3} = 12.7$$ ***miles, the approximate height of the tower***

The length of the hypotenuse is double the length of the short leg:

$$2(22/\sqrt{3}) \approx 25.4$$ ***miles, the distance the signal travels***

5. The measure of the longest leg in this 30-60-90 triangle is 42 miles. Find the length of the hypotenuse.

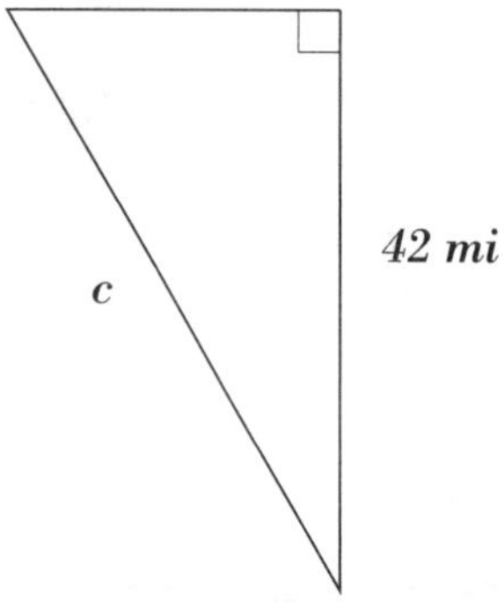

The longest leg is the side opposite the 60° angle. The length of the side opposite the 60° angle is half the length of the hypotenuse (the short leg) times $\sqrt{3}$:

$$42 = \frac{1}{2} \bullet c\sqrt{3}$$

$$c = \frac{84}{\sqrt{3}} \approx 48.5 \text{ miles}$$

6. The length of the shortest leg in this 30-60-90 triangle is 33 meters. Find the length of the hypotenuse.

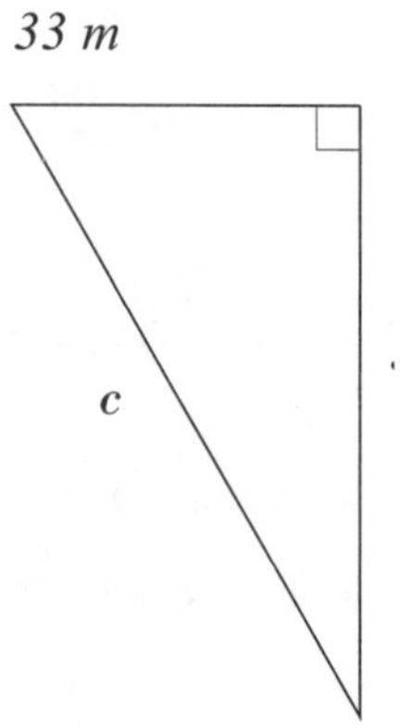

The shortest leg is the side opposite the 30° angle. The length of the side opposite the 30° angle is half the length of the hypotenuse:

$$33 = \frac{1}{2} \bullet c$$

$$c = 66 \text{ m}$$

Special Right Triangle Quiz 2 (Form B)

Name: ______________________________

Show the work required to solve each problem.

1. If the measure of the longest leg in this 30-60-90 triangle is 10 meters, find the length of sides *a* and *c*.

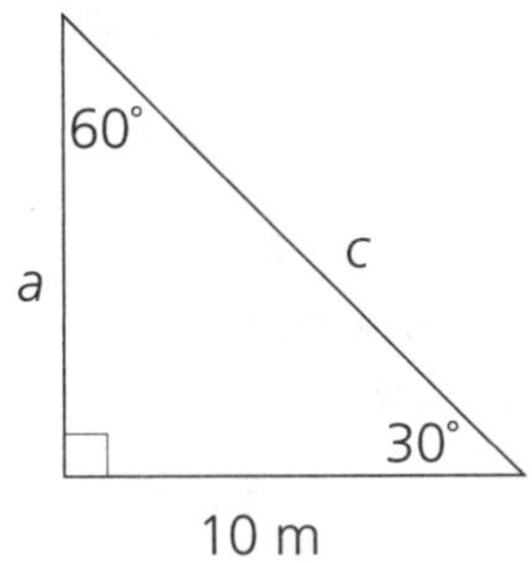

The length of the side opposite the 30° angle is the length of the side opposite the 60° angle divided by the $\sqrt{3}$:

$$a = 10/\sqrt{3} \approx 5.77 \text{ m (approximately)}$$

The length of the hypotenuse is double the length of the short leg:

$$2(10/\sqrt{3}) \approx 11.55 \text{ m}$$

2. If the measure of the shortest leg of this 30-60-90 triangle is 25 feet, find the length of sides *b* and *c*.

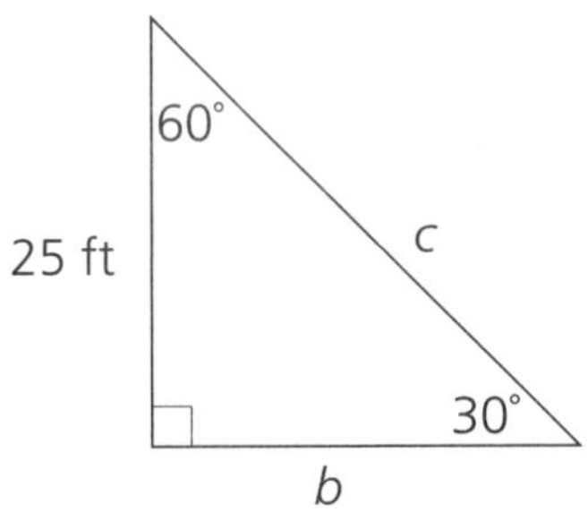

The length of the hypotenuse is double the length of the side opposite the 30° angle:

$$2 \bullet 25 = c$$

$$c = 50 \text{ ft}$$

The length of the side opposite the 60° angle is the length of the short leg times $\sqrt{3}$:

$$b = 25\sqrt{3} \approx 43.3 \text{ ft}$$

3. If the measure of the hypotenuse in this 30-60-90 triangle is 9 mm, find the length of sides *a* and *b*.

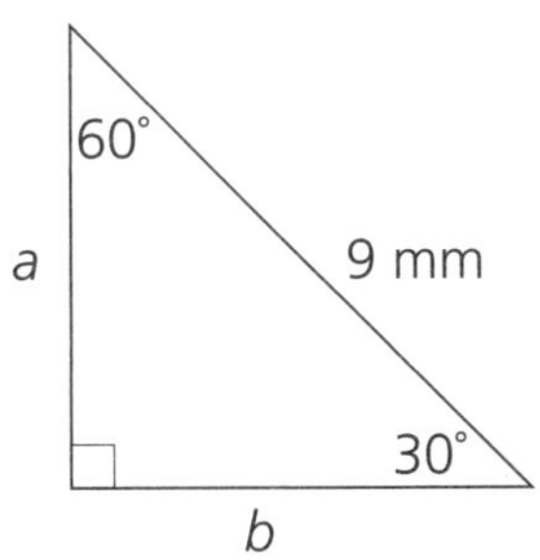

The length of the side opposite the 30° angle is half the length of the hypotenuse:

$$a = \frac{1}{2} \bullet 9 = 4.5 \text{ mm}$$

The length of the side opposite the 60° angle is the length of the short leg times $\sqrt{3}$:

$$b = 4.5\sqrt{3} \approx 7.79 \text{ mm}$$

4. A microwave dish is situated on top of a 50-foot tower; the signal travels from the dish to your house so you can watch cable TV. The angle of elevation from your house to the dish measures 30°. Find the distance from your house to the base of the tower and the distance the signal travels.

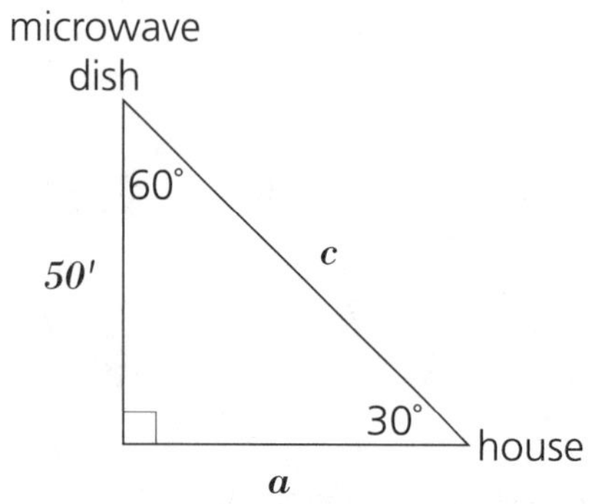

The length of the side opposite the 30° angle is half the length of the hypotenuse:

$$50 = \frac{1}{2} \bullet c$$

c = 100 ft, the distance the signal travels

The length of the side opposite the 60° angle is the length of the short leg times $\sqrt{3}$:

$a = 50\sqrt{3} \approx 86.6$ ft, the distance from your house to the tower

5. The measure of the shortest leg in this 30-60-90 triangle is 42 meters. Find the length of the hypotenuse.

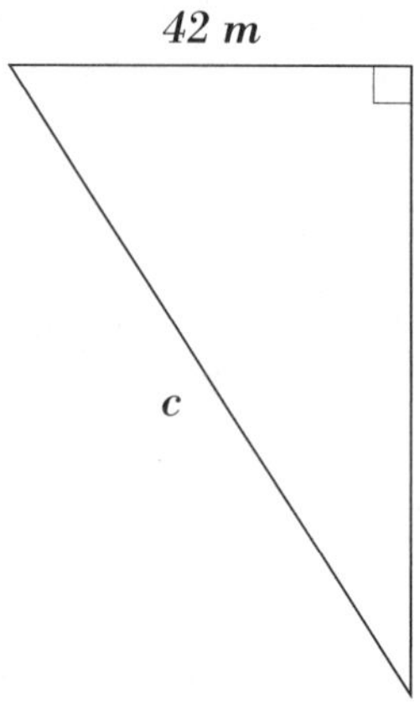

The shortest leg is the side opposite the 30° angle. The length of the side opposite the 30° angle is half the length of the hypotenuse:

$$42 = \frac{1}{2} \bullet c$$

$$c = 84 \text{ m}$$

6. The measure of the longest leg in this 30-60-90 triangle is 33 meters. Find the length of the hypotenuse.

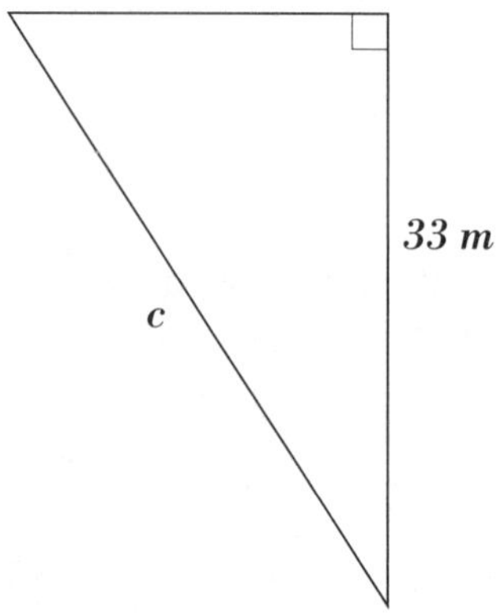

The longest leg is the side opposite the 60° angle. The length of the side opposite the 60° angle is half the length of the hypotenuse times $\sqrt{3}$:

$$33 = \frac{1}{2} \bullet c\sqrt{3}$$

$$c = \frac{66}{\sqrt{3}} \text{ m} \approx 38.11 \text{ m}$$

Special Right Triangle Quiz 2 (Form C)

Name: ______________________

Show the work required to solve each problem.

1. If the area of this trapezoid is 156 cm^2, find its perimeter.

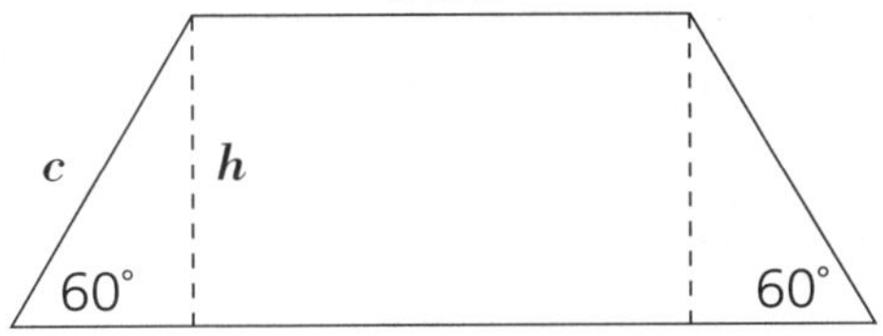

$$A = \frac{1}{2}h(b_1 + b_2)$$

$$156 = \frac{1}{2} \bullet h \bullet (8 + 16)$$

$$156 = 12h$$

$$h = 13 \text{ cm}$$

The longest leg is the side opposite the 60° angle. The length of the side opposite the 60° angle is half the length of the hypotenuse times $\sqrt{3}$:

$$13 = \frac{1}{2} \bullet c\sqrt{3}$$

$$c = \frac{26}{\sqrt{3}} \approx 15.01 \text{ cm}$$

Perimeter of the trapezoid:

$$P = 15.01 + 8 + 15.01 + 16 = 54.02 \text{ cm}$$

2. Find the area of this triangle.

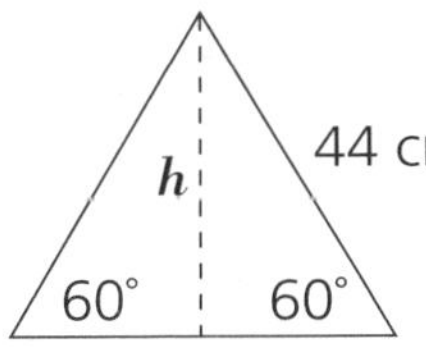

Draw a height segment as shown. This height is the side opposite the 60° angle in each 30-60-90 triangle.

The length of the side opposite the 60° angle is half the length of the hypotenuse times $\sqrt{3}$:

$$h = \frac{1}{2} \bullet 44\sqrt{3} = 22\sqrt{3} \approx 38.11 \text{ cm}$$

Area of the original triangle:

$$A = \frac{1}{2}bh = \frac{1}{2} \bullet 44 \bullet 38.11 = 838.42 \text{ cm}^2 \text{ (approximately)}$$

3. Find the area of this trapezoid.

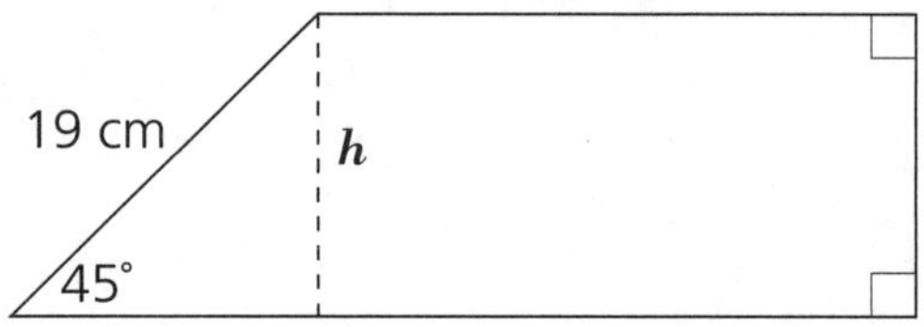

Draw a height segment as shown. This height segment is a leg of the 45-45-90 triangle, in which the hypotenuse is the length of a leg times $\sqrt{2}$:

$$\text{hypotenuse} = \text{leg} \bullet \sqrt{2}$$

$$19 = \text{leg} \bullet \sqrt{2}$$

$$\text{leg} = \frac{19}{\sqrt{2}} \approx 13.44 \text{ cm}$$

The longer base of the trapezoid is approximately 13.44 + 22, or 35.44 cm.

Area of the trapezoid:

$$A = \frac{1}{2}h(b_1 + b_2)$$

$$A \approx \frac{1}{2} \bullet 13.44 \bullet (35.44 + 22)$$

$$A \approx 386 \text{ cm}^2$$

Special Right Triangle Quiz 2 (Form D)

Name: ______________________

Show the work required to solve each problem.

1. If the area of this trapezoid is 130 cm², find its perimeter.

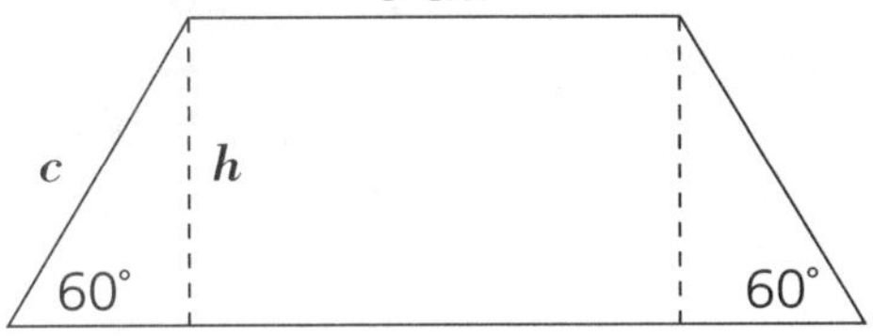

$$A = \frac{1}{2}h(b_1 + b_2)$$

$$130 = \frac{1}{2} \bullet h \bullet (8 + 16)$$

$$130 = 12h$$

$$h \approx 10.83 \text{ cm}$$

The longest leg is the side opposite the 60° angle. The length of the side opposite the 60° angle is half the length of the hypotenuse times $\sqrt{3}$:

$$10.83 \approx \frac{1}{2} \bullet c\sqrt{3}$$

$$c \approx \frac{21.66}{\sqrt{3}} \approx 12.51 \text{ cm}$$

Perimeter of the trapezoid:

$$P \approx 12.51 + 8 + 12.51 + 16 = 49.02 \text{ cm}$$

2. Find the area of this triangle.

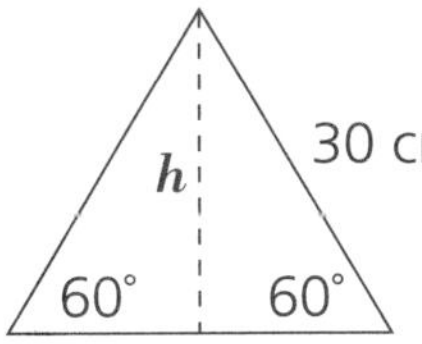

Draw a height segment as shown. This height is the side opposite the 60° angle in each 30-60-90 triangle.

The length of the side opposite the 60° angle is half the length of the hypotenuse times $\sqrt{3}$:

$$h = \frac{1}{2} \bullet 30\sqrt{3} = 15\sqrt{3} \approx 25.98 \text{ cm}$$

Area of the original triangle:

$$A = \frac{1}{2}bh = \frac{1}{2} \bullet 30 \bullet 25.98 = 389.7 \text{ cm}^2 \text{ (approximately)}$$

3. Find the area of this trapezoid.

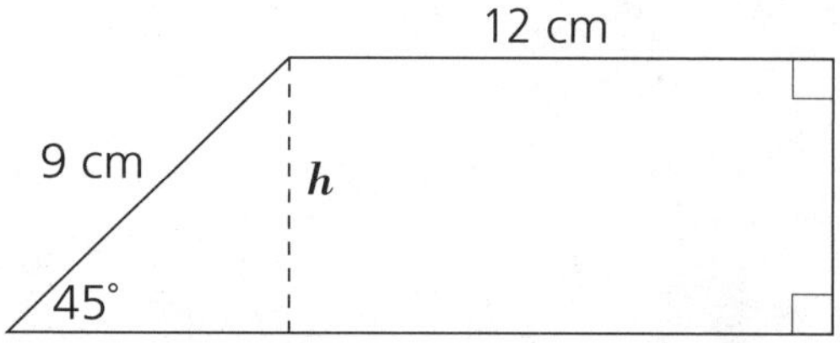

Draw a height segment as shown. This height segment is a leg of the 45-45-90 triangle, in which the hypotenuse is the length of a leg times $\sqrt{2}$:

$$\textit{hypotenuse} = \textit{leg} \bullet \sqrt{2}$$

$$9 = \textit{leg} \bullet \sqrt{2}$$

$$\textit{leg} = \frac{9}{\sqrt{2}} \approx 6.36 \text{ cm}$$

The longer base of the trapezoid is approximately 6.36 + 12, or 18.36 cm.
Area of the trapezoid:

$$A = \frac{1}{2}h(b_1 + b_2)$$

$$A \approx \frac{1}{2} \bullet 6.36 \bullet (18.36 + 12)$$

$$A \approx 96.54 \text{ cm}^2$$

Quiz: Right Triangle Trigonometry 1

Name: ____________________________

The following statement is the ADA Title III requirement for wheelchair ramps.

The least possible slope shall be used for any ramp. The maximum slope of a ramp in new construction shall be 1:12. The maximum rise for any run shall be 30 inches. —Title III, Section 4.8.2

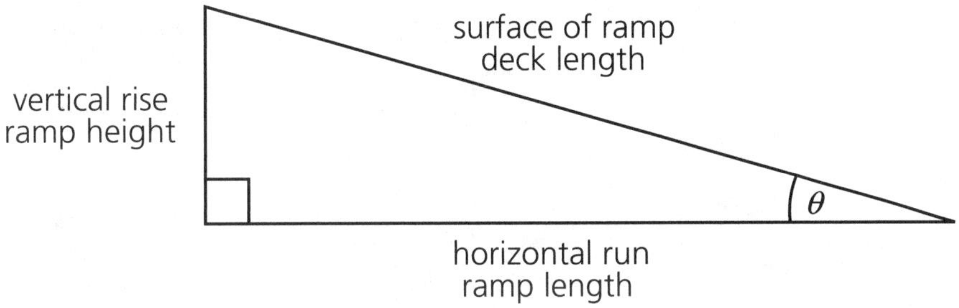

1. According to the law, the slope (the ratio of the vertical rise to the horizontal run) must be no greater than 1:12. This means the maximum ratio of the ramp height to the ramp length is 1:12.

   Complete the following table so the ratio of the height to the length always meets this criterion. Use the Pythagorean theorem to find the deck length.

Height	Length	$\frac{\text{Height}}{\text{Length}}$	Deck Length
3 feet	***36 feet***	***1:12***	***36.12 feet***
1.67 feet	20 feet	***1:12***	***20.07 feet***
0.42 meters	5 meters	***1:12***	***5.02 meters***
1 meter	***12 meters***	***1:12***	***12.04 meters***
x meters	***12x meters***	***1:12***	***12.04x meters***
$\frac{y}{12}$ ***feet***	*y* feet	***1:12***	***1.003y feet***
26 inches	***312 inches***	***1:12***	***313.1 inches***

2. Complete the following table in a similar manner. This time, however, use a height to length ratio of 1:15.

Height	Length	Deck Length
36 inches	***540 inches***	***541.20 inches***
0.8 feet	12 feet	***12.03 feet***
0.33 meters	5 meters	***5.01 meters***
1.5 meters	***22.5 meters***	***22.55 meters***
x meters	***15x meters***	***15.03x meters***
$\frac{y}{15}$ ***feet***	*y* feet	***1.002y feet***
24 inches	***360 inches***	***360.8 inches***

Quiz: Right Triangle Trigonometry 2

Name: ______________________________

1. You are standing a certain distance from the base of the Washington Monument. The angle of elevation from your feet to the top of the monument is 20°. You know from your tour book that the height of the monument is 555 feet. How far are you from the base of the monument?

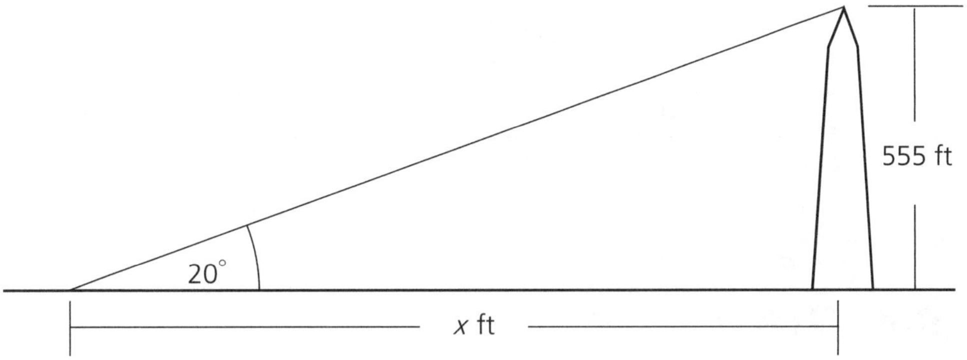

$$\tan 20° = \frac{555\text{ ft}}{x\text{ ft}}$$

$$x\text{ feet} = \frac{555\text{ ft}}{\tan 20°}$$

$$x = 1524.85\text{ feet}$$

2. A pilot approaches an airport for a landing. The pilot is 50 km from the airport and measures the angle of depression to the airport to be 15°. What is the pilot's altitude?

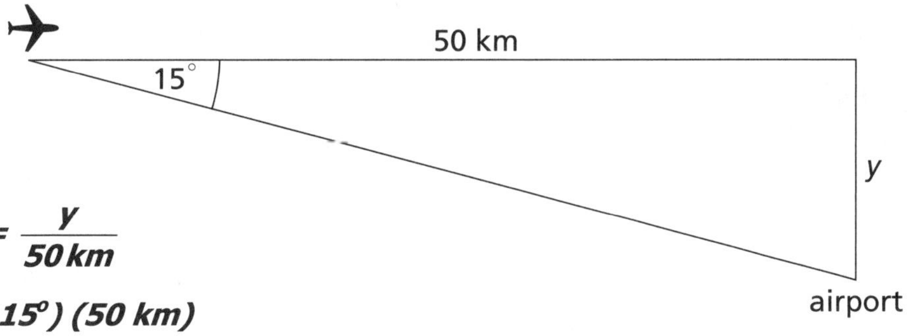

$$\tan 15° = \frac{y}{50\text{ km}}$$

$$y = (\tan 15°)(50\text{ km})$$

$$y = 13.4\text{ km}$$

3. A firefighter is called to rescue a cat caught in a tree. If the firefighter is standing 30 m from the base of the tree and if the angle of elevation to the cat is 20°, how high in the tree is the cat?

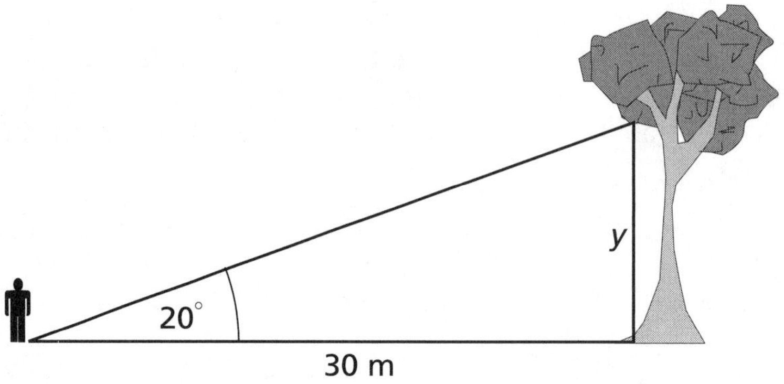

$$\tan 20^\circ = \frac{y}{30\ m}$$

$$y = (\tan 20^\circ)(30\ m)$$

$$y = 10.9\ m$$

4. You are a camera operator in a blimp over a stadium. Your blimp is 200 m directly over one end zone. The distance between the end zones is 100 m. What angle of depression will allow you to film the action in the *other* end zone?

$$\tan\theta = \frac{200\ m}{100\ m}$$

$$\theta = \tan^{-1} \frac{200\ m}{100\ m}$$

$$\theta = 63.4^\circ$$

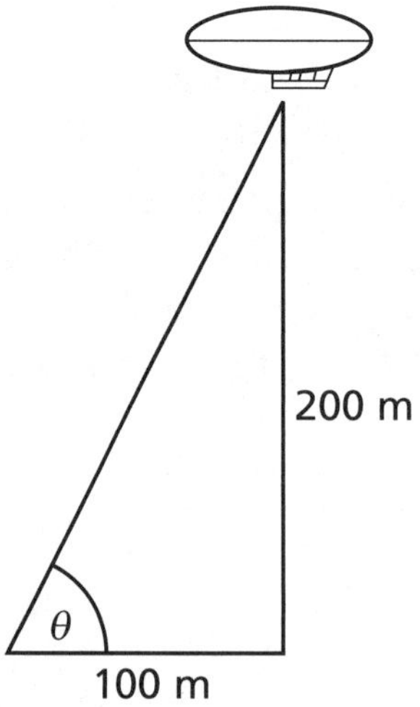

Squares: Part 1

Name: ____________________________

For each equation:

- Enter the equation into your graphing calculator and graph it
- Fill in at least five *x* and *y* values in the table
- Sketch the graph
- Find the *x*-intercepts and the *y*-intercept
- Find the equation for the axis of symmetry
- Find the coordinates of the vertex

1. $y = x^2 + 5x + 4 =$ (***x + 4***)(***x + 1***)

x	y
0	***4***
-4	***0***
-1	***0***
-2.5	***-2.25***
-5	***4***

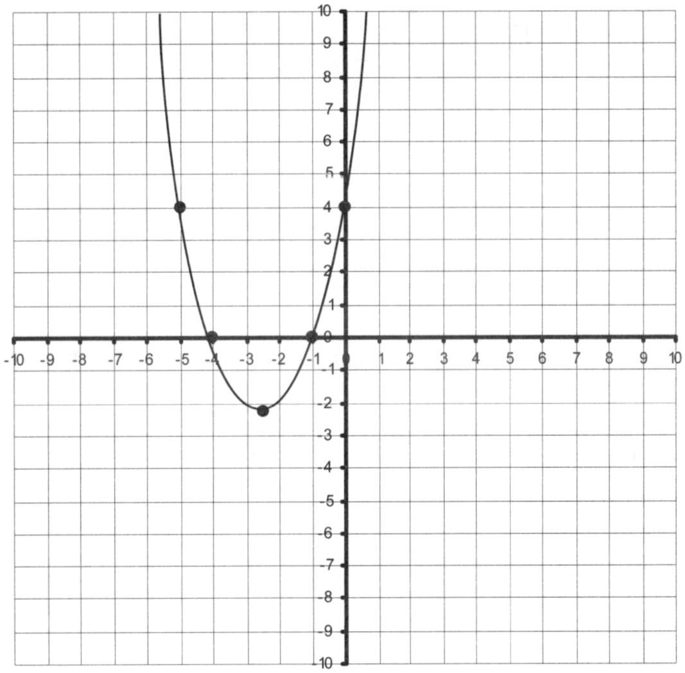

x-intercepts: ***(-4, 0) (-1, 0)***

y-intercept: ***(0, 4)***

Axis of symmetry: *x* = ***-2.5***

Vertex: (***-2.5*** , ***-2.25***)

2. $y = x^2 - 5x + 6 =$ **(x − 3)(x − 2)**

x	y
0	***6***
3	***0***
2	***0***
2.5	***-0.25***
5	***6***

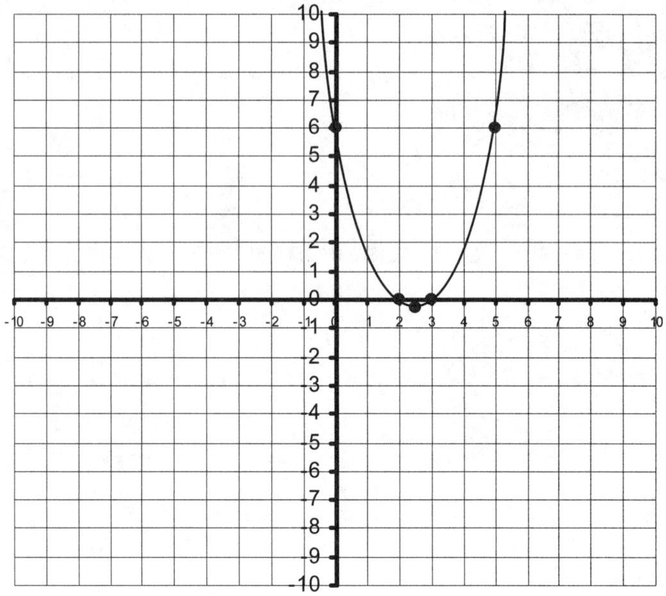

x-intercepts: ***(3, 0) (2, 0)***

y-intercept: ***(0, 6)***

Axis of symmetry: *x* = ***(2.5)***

Vertex: (***2.5*** , ***-0.25***)

Squares: Part 2

Name: ____________________________

For each equation:

- Enter the equation into your graphing calculator and graph it
- Fill in at least five x and y values in the table
- Sketch the graph
- Find the x-intercepts and the y-intercept
- Find the equation for the axis of symmetry
- Find the coordinates of the vertex

1. $y = x^2 + 7x + 6 =$ (***x + 1***)(***x + 6***)

x	y
0	*6*
-1	*0*
-6	*0*
-3.5	*-6.25*
-7	*6*

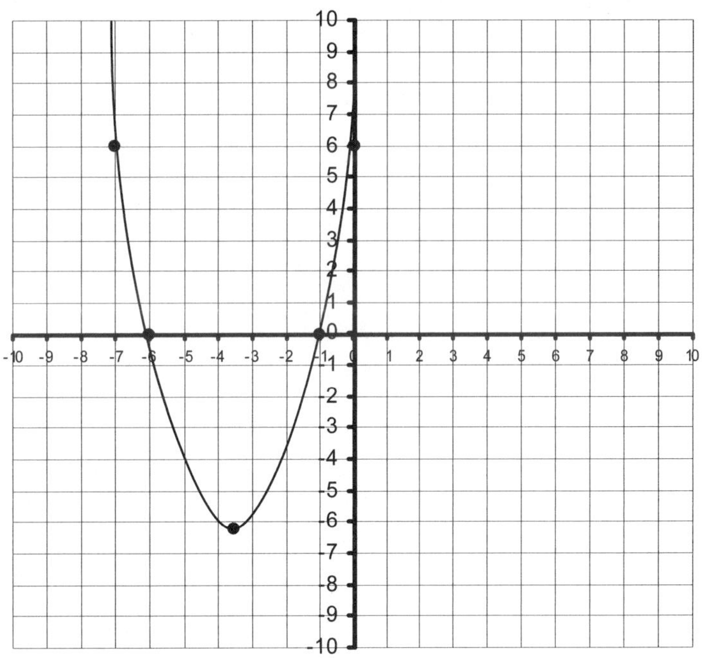

x-intercepts: ***(-1, 0) (-6, 0)***

y-intercept: ***(0, 6)***

Axis of symmetry: x = ***-3.5***

Vertex: (***-3.5*** , ***-6.25***)

2. $y = x^2 - 6x + 8 =$ (***x − 2***)(***x − 4***)

x	y
0	***8***
2	***0***
4	***0***
6	***8***
3	***-1***

x-intercepts: ***(2, 0) (4, 0)***

y-intercept: ***(0, 8)***

Axis of symmetry: $x =$ ***3***

Vertex: (***3*** , ***-1***)

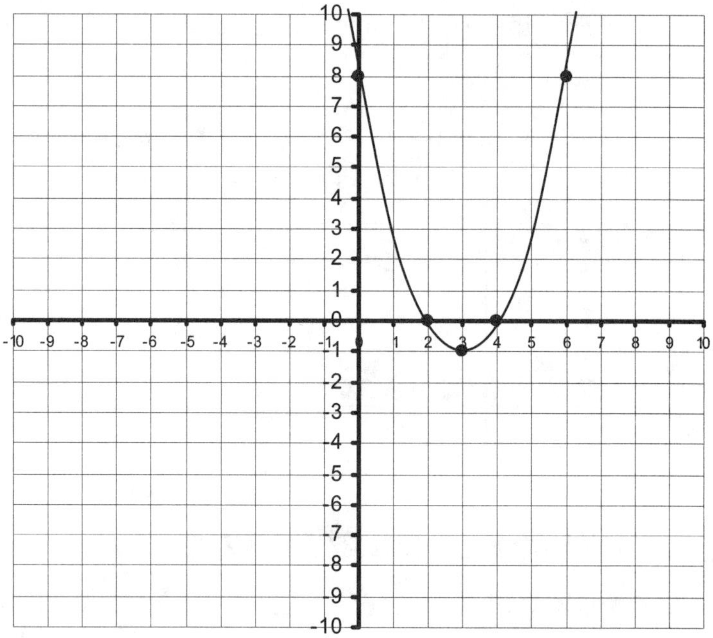

Quadratic Formula

Name: ______________________

General form of the quadratic equation:

$$y = Ax^2 + Bx + C$$

Quadratic formula:

$$x = \frac{-B \pm \sqrt{B^2 - 4AC}}{2A}$$

For each quadratic equation:

- Write the values of A, B, and C from the general form
- Use the quadratic formula to solve the equation for the x-intercepts
- Use the x-intercepts to write the equation in factored form

1. $y = x^2 - 9x - 10 = (\ \boldsymbol{x + 1}\ )(\ \boldsymbol{x - 10}\ )$

   $A =$ **1** $B =$ **-9** $C =$ **-10**

   x = 10 **x = -1**

   x-intercepts: **(10, 0) (-1, 0)**

2. $y = 2x^2 - 16x - 128 = (\quad)(\quad)$ ***No integer factors.***

   $A =$ **2** $B =$ **-16** $C =$ **-128**

   x = 12.9443 **x = -4.9443**

   x-intercepts: **(12.9443, 0) (-4.9443, 0)**

3. $y = 3x^2 + 8x - 10 = (\quad)(\quad)$ ***No integer factors.***

   $A =$ **3** $B =$ **8** $C =$ **-10**

   x = 0.9274 **x = -3.5941**

   x-intercepts: **(0.9274, 0) (-3.5941, 0)**

Vertical Motion

Name: ______________________________

A cannonball is shot straight up into the air from the side of a mountain. The flight of the cannonball can be modeled by this equation for vertical motion:

$$y = -16x^2 + vx + h$$

In this equation, y is the number of feet above the valley, x is the time in seconds from the moment the ball is shot, v is the initial velocity (or speed) in feet per second, and h is the initial height in feet of the cannonball above the valley floor.

The initial velocity of the cannonball is 600 feet per second, and the cannon is 100 feet above the valley floor.

1. Write the equation for flight of the cannonball.

   $y = -16x^2 + 600x + 100$

Use complete sentences to answer the following questions.

2. How high is the cannonball:

   a. 1 second *after* it is shot?

      684 feet

   b. 20 seconds *after*?

      5,700 feet

   c. 50 seconds *after*?

      -9,900 feet

   d. 5 seconds *before* it is shot?

      -3,300 feet

   e. 50 seconds *before* it is shot?

      -69,900 feet

3. Do all the answers to question 2 make sense in this problem situation?

   No, all of these responses do not make sense. The times in (d) and (e) before the cannonball was shot do not make sense. Before the ball was shot, it was not in motion. Also (c) does not make sense since the cannonball was shot 100 feet above the floor of the valley.

4. Use the quadratic formula to find when the cannonball is on the ground. How do you know?

   -0.1659 seconds or 37.6659 seconds

5. When is the cannonball at its highest point? How high does it get?

   18.75 seconds or 5,725 feet

6. Use the quadratic formula to find when the cannonball will be back at its original level. How do you know?

$$-16x^2 + 600x + 100 = 100$$

$$x = \frac{-600 \pm \sqrt{600^2 - 4(-16)(0)}}{2(-16)}$$

$$x = \frac{-600 + 600}{-32}$$

$$x = \frac{0}{-32}$$

Answer: 0 seconds

$$-16x^2 + 600 = 0$$

$$x = \frac{-600 \pm \sqrt{600^2 - 32}}{-32}$$

$$x = \frac{-600 - 600}{-32}$$

$$x = \frac{-1200}{-32}$$

Answer: 37.5 seconds

7. When is the cannonball 3000 feet above the valley floor?

$$-16x^2 + 600x + 100 = 3000$$

$$x = \frac{-600 \pm \sqrt{600^2 - 4(-16)(-2900)}}{2(-16)}$$

$$x = \frac{-600 + \sqrt{17440}}{-32}$$

$$x = \frac{-600 + 417.6123}{-32}$$

$$x = \frac{-182.3877}{-32} \approx 5.7$$

$$-16x^2 + 600x - 2900 = 0$$

$$x = \frac{-600 \pm \sqrt{360000 - 185600}}{-32}$$

$$x = \frac{-600 - \sqrt{174,400}}{-32}$$

$$x = \frac{-600 - 417.6123}{-32}$$

$$x = \frac{-1017.6123}{-32} \approx 31.8$$

8. Use the information you have found so far about the problem situation to fill in this table.

Labels	***Time***	***Height above Valley***
Units	***seconds***	***feet***
Formula	x	$-16x^2 + 600x + 100$
	1	***684***
	20	***5,700***
	18.75	***5,725***
	0	***100***
	37.5	***100***

9. Graph your equation on the following grid.

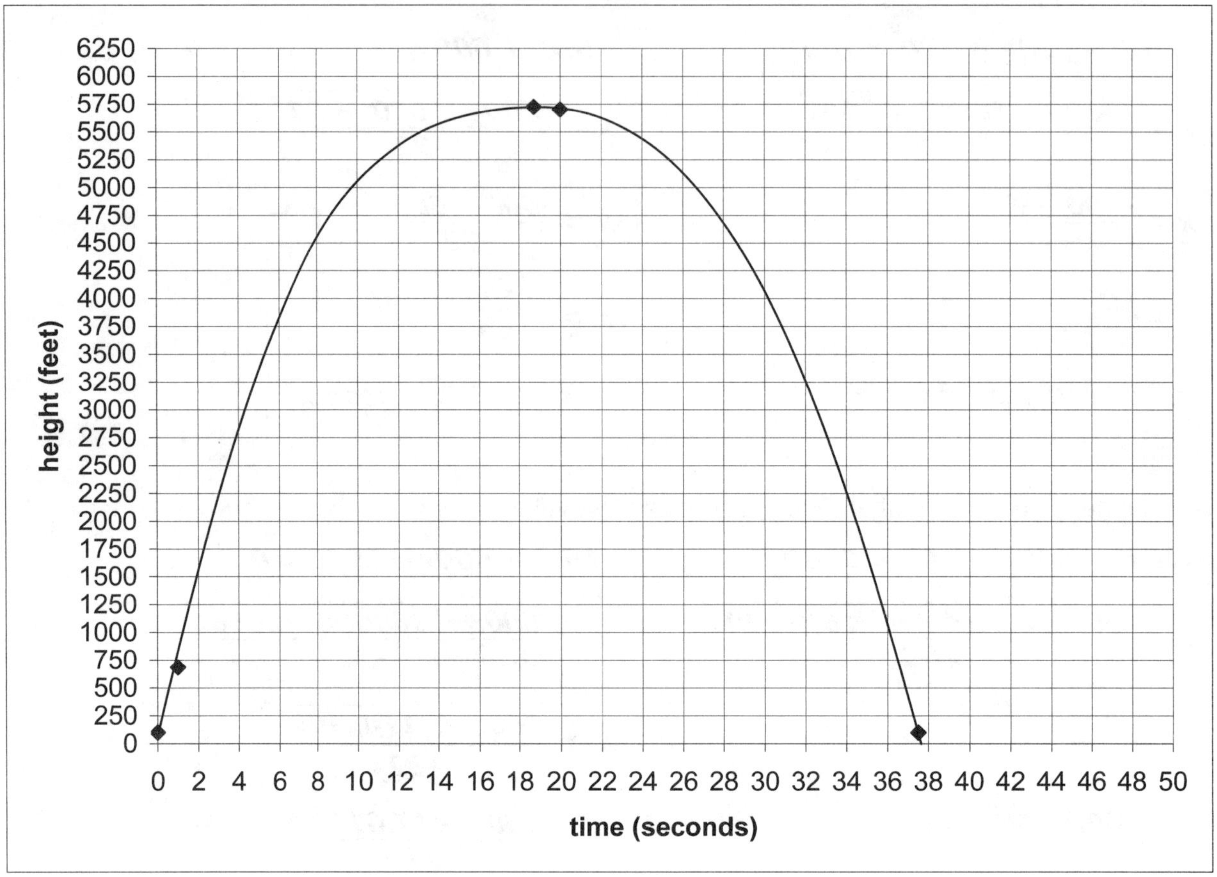

Data

Name: ____________________________

1. Use the data table to graph the points. Then find the slope, the *y*-intercept, and the equation that *fits* the data points.

x	y
1	10
4	31
3	24
8	51
11	63

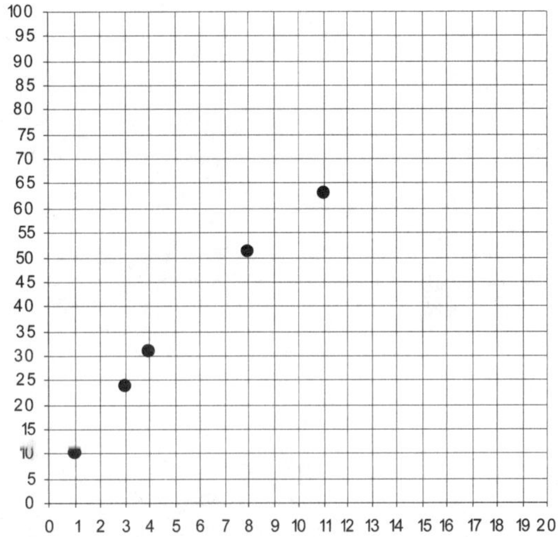

Slope: **5.22**

y-intercept: **(0, 7.61)**

Equation: $y =$ **5.22x + 7.61**

For each set of data, calculate the mean, the median, and the mode.

2. Data: 14, 23, 24, 19, 15, 18, 19

mean: $\bar{x} = \frac{132}{7} = 18\frac{6}{7} \approx 18.86$ median = **19** mode = **19**

3. Data: 1111, 1415, 1612, 1111, 1313

mean: $\bar{x} = \frac{6562}{5} = 1312\frac{2}{5} = 1312.4$ median = **1313** mode = **1111**

Running the Mile

Name: ____________________________

This table shows the fastest time in seconds that a human ran the mile from 1945 to 1985. Find the slope, the y-intercept, and the equation of the linear regression line for *number of years since 1945* versus *time in seconds*. Then answer the questions.

Year	***Time in Seconds***
1945	*241.4*
1954	*238.0*
1957	*237.2*
1962	*234.4*
1964	*234.1*
1965	*233.6*
1966	*231.3*
1967	*231.1*
1975	*229.4*
1979	*229.0*
1980	*228.8*
1981	*228.4*
1981	*227.3*
1985	*226.3*

Slope: b = ***-0.3674*** y-intercept: a = ***(0, 240.8495)***

Equation: y = ***-0.3674x + 240.8495*** Regression coefficient: r = ***-0.9818***

Write your answers in complete sentences.

1. Use your model (equation) to predict how fast a runner in 1990 should be able to run the mile.

45 years • -0.3674 = -16.533

-16.533 + 240.8495 = 224.3165

Answer: 224.3165 seconds

2. Use your model (equation) to predict how fast a runner in 1993 should be able to run the mile.

48 years • -0.3674 = -17.6352

-17.6352 + 240.8495 = 223.2143

Answer: 223.2143 seconds

3. In what year does your model (equation) predict that the fastest time a runner ran the mile was 237.5 seconds?

-0.3674x + 240.8495 = 237.5	***1945***
-0.3674x = -3.3495	***+ 9***
x = 9.1168 Answer:	***1954***
x = 9 years	

4. In what year does your model (equation) predict that 200 seconds is the fastest time a runner will run the mile?

-0.3674x + 240.8495 = 200	***1945***
-0.3674x = -40.8495	***+ 111***
x = 111.1854 Answer:	***2056***
x = 111 years	

5. How good a fit is your model? Explain your thinking.

The model is a good fit because the regression coefficient value is close to –1.

Probability on the Shuffleboard Court

Name: ___________________________

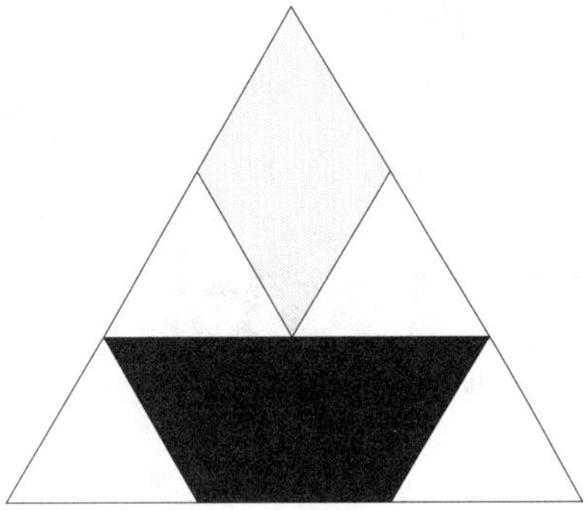

1. Your friend Sasha calls you in the evening while you are on the phone with another friend, Jordan. Sasha is confused about something in his class notes. You decide to conference him so he can also talk to Jordan. Sasha said that he wrote the probability of landing on the trapezoid on the shuffleboard triangle pictured above is 1/3, but without understanding why. He tells you he thinks this is so because on the shuffleboard, he sees three different shapes: a triangle, a rhombus and a trapezoid. Jordan jumps in saying that that reason is wrong since he thinks there are four triangles, a rhombus and a trapezoid there are six shapes, so that the probability of landing on the trapezoid is 1/6, not 1/3. You disagree with Jordan's answer since you have the same answer as Sasha, however you disagree with Sasha's reasoning. Explain your reasoning.

   The reason the probability of landing on the trapezoid is 1/3 is because the trapezoid accounts for 1/3 of the area of the shuffleboard triangle, not because it is 1 of 3 shapes.

2. How is area related to probability in this problem?

   The fraction of area accounted for by each shape is the probability of landing on that shape.

3. Find the probability of landing on the shaded triangle embedded in this square.

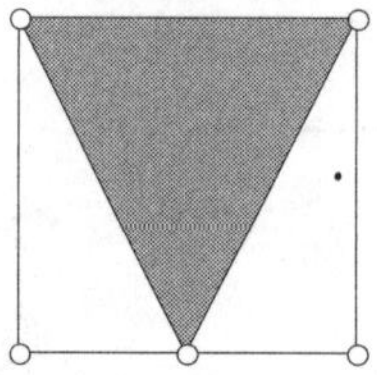

2 cm

Area of the square = 4 cm^2

Area of the shaded triangle = ½ 4 = 2 cm^2

Probability of landing on the shaded region = $\frac{2}{4} = \frac{1}{2}$